TEACHING CONTENT MANAGEMENT IN TECHNICAL AND PROFESSIONAL COMMUNICATION

This collection offers a comprehensive overview of approaches to teaching the complex subject of content management.

The 12 chapters define and explain content management and its accompanying competencies, providing teaching examples in areas including content strategy, topic-based writing, usability studies, and social media. The book covers tasks associated with content management such as analyzing audiences and using information architecture languages including XML and DITA. It highlights the communal aspects of content management, focusing on the work of writing stewardship and project management, and the characteristics of content management in global contexts. It concludes with a look to the future and the forces that shape content management today. The editor situates the collection within a pedagogical exigency, providing sound instructional approaches to teaching content management from a rhetorical perspective.

The book is an essential resource for both instructors new to teaching technical and professional communication, and experienced instructors who are interested in upgrading their pedagogies to include content management.

Tracy Bridgeford is a professor of technical communication at the University of Nebraska at Omaha, where she directs the Graduate Certificate in Technical Communication. In 2018, she published *Teaching Professional and Technical Communication: A Practicum in a Book*. She coedited *Academy-Industry Relationships: Perspectives for Technical Communicators*, *Sharing Our Intellectual Traces: Narrative Reflections from Administrators of Professional, Technical, and Scientific Programs*, and *Innovative Approaches to Teaching Technical Communication*. She contributed chapters to *Editing in the Modern Classroom*, *Resources in Technical Communication: Outcomes and Approaches*, and *Teaching Writing with Computers: An Introduction*.

ATTW Book Series in Technical and Professional Communication

Tharon Howard, Series Editor

Rhetorical Work in Emergency Medical Services: Communicating in the Unpredictable Workplace
Elizabeth L. Angeli

Involving the Audience: A Rhetorical Perspective on Using Social Media to Improve Websites
Lee-Ann Kastman Breuch

Creating Intelligent Content with Lightweight DITA
Carlos Evia

Editing in the Modern Classroom
Michael J. Albers and Suzan Flanagan

Translation and Localization: A Guide for Technical Communicators
Bruce Maylath, and Kirk St.Amant

Technical Communication After the Social Justice Turn: Building Coalitions for Action
Rebecca Walton, Kristen R. Moore and Natasha N. Jones

Content Strategy in Technical Communication
Guiseppe Getto, Jack T. Labriola and Sheryl Ruszkiewicz

Teaching Content Management in Technical and Professional Communication
Tracy Bridgeford

For additional information on this series please visit www.routledge.com/ ATTW-Series-in-Technical-and-Professional-Communication/book-series/ATTW, and for information on other Routledge titles visit www.routledge.com.

TEACHING CONTENT MANAGEMENT IN TECHNICAL AND PROFESSIONAL COMMUNICATION

Edited by Tracy Bridgeford

NEW YORK AND LONDON

First published 2020
by Routledge
52 Vanderbilt Avenue, New York, NY 10017

and by Routledge
2 Park Square, Milton Park, Abingdon, Oxon, OX14 4RN

Routledge is an imprint of the Taylor & Francis Group, an informa business

Library of Congress Cataloging-in-Publication Data
Names: Bridgeford, Tracy, 1960- editor.
Title: Teaching content management in technical and professional communication / edited by Tracy Bridgeford.
Description: New York, NY : Routledge, 2020. |
Series: ATTW book series in technical and professional communication | Includes bibliographical references and index.
Identifiers: LCCN 2019045374 (print) | LCCN 2019045375 (ebook) |
ISBN 9780367181253 (hardback) | ISBN 9780367181260 (paperback) |
ISBN 9780429059612 (ebook)
Subjects: LCSH: Communication of technical information–Study and teaching.
Classification: LCC T10.5 .T327 2020 (print) | LCC T10.5 (ebook) |
DDC 607.1–dc23
LC record available at https://lccn.loc.gov/2019045374
LC ebook record available at https://lccn.loc.gov/2019045375

ISBN: 978-0-367-18125-3 (hbk)
ISBN: 978-0-367-18126-0 (pbk)
ISBN: 978-0-429-05961-2 (ebk)

Typeset in Bembo
by Swales & Willis, Exeter, Devon, UK

CONTENTS

ILLUSTRATIONS

Figures

Tables

Boxes

SERIES EDITOR FOREWORD

Tracy Bridgeford's edited collection *Teaching Content Management in Technical and Professional Communication* is a most welcome addition to the ATTW Book Series in Technical and Professional Communication (TPC) because, as its title suggests, it is a much needed book about how we, as teachers of technical writing, must prepare students for a workplace which is now dominated by technologies such as Extensible Markup Language (XML), single-source authoring environments, usability testing, and, most particularly, content management systems (CMSs). Like all the other books in the ATTW Book Series, *Teaching Content Management in Technical and Professional Communication* is solidly based on the expertise of the best and most experienced educators in our field. In this single volume, Tracy has assembled a literal *Who's Who* of scholars and educators who speak authoritatively to the question of how to teach CMSs from perspectives ranging from usability studies, to content strategy, to XML/DITA (Darwin Information Typing Architecture) authoring, to social media management, to content reuse. No matter how you might plan to approach CMSs in your classroom, you will find multiple perspectives on it written by the leading scholars in TPC.

For readers who are unfamiliar with CMSs, Bridgeford's work is firmly situated in what is starting to be called the shift from the "craftsman model" of technical writing to what Rebekka Anderson (who contributed to the Afterword in this volume) has called the "Component Content Management model." This shift has come about as a result of the ways that new technologies like XML/DITA, single-source authoring environments, and CMSs have changed the ways that technical writers deliver information to audiences (or what we would now call "users"). As Carlos Evia, another author in the ATTW Series, wrote in his book *Creating Intelligent Content with Lightweight*

DITA, the "craftsman" model made technical communicators responsible for the development and delivery of *entire* documents—i.e., complete documentation sets, recommendation reports, marketing brochures, etc. And as Mike Albers has observed, our curricula and textbooks are still based on the assumption that our students are being prepared as craft persons who will do this same sort of work in their careers. The Component Content Management paradigm shift—which Tatiana Batova and Rebekka Andersen outline in their 2017 *IEEE Transactions* article on skills needed for Component Content Management, and which JoAnn Hackos describes in her 2015 ISO/IEC/IEEE 26531 standard on content management—has made clear that today our graduates don't "write" so much as they develop, delineate, and manage small content modules which they collect into information architectures. Today's technical communicators must not only write well, but now they also need to pull on a knowledge of content strategy, information architectures, user experience design, and single-source authoring environments in order to produce usable "texts" which are delivered on their company's CMS.

This book is an extremely timely and much needed introduction to the Component Content Management and the computational thinking movement in TPC. As such, it's a pleasure to have it in the ATTW Book Series in Technical and Professional Communication.

Dr. Tharon W. Howard
Editor, ATTW Book Series in
Technical and Professional Communication
October 17, 2019

ACKNOWLEDGMENTS

I wish to thank Kathy Radosta and Joan Latchaw. A special thank you to Bill Williamson for his continued support and friendship and wisdom.

INTRODUCTION

Content Management: A Pedagogical Exigency

Tracy Bridgeford

UNIVERSITY OF NEBRASKA AT OMAHA

Chapter Takeaways

- Because of the existence of content management systems (CMSs), we are in the midst of a pedagogical exigency.
- Three conversations record the changes wrought by content management systems: single-sourcing methodologies, content management, and the craft tradition.
- Craft is not dead as a result of these changes. It needs to be redefined as networked agency.
- Networked agency can help us move from the collaborative model of collaboration to a shared idea of writing.

More than a decade ago, George Pullman and Baotong Gu (2008) framed an exigence in their introduction to a special issue of *Technical Communication Quarterly* that "it is high time for our field not only to gain a better understanding of CMSs" but also to formulate "theoretically sound and pedagogically viable" approaches to content management (p. 3; see also Pullman & Gu, 2009). Although they were absolutely correct, our pedagogical community has not responded with any sense of urgency. It is imperative that we provide critical, sustainable approaches to teaching content management and strategy that emphasize the rhetorical value of content. All teachers of professional and technical communication share the pedagogical burden of responding to this challenge. We must help students learn to reflect on the design process, to strategize about content development, and to collaborate in networks of content creators and strategists to meet the needs of local and global users. This collection responds to the exigence by gathering resources and strategies from the teaching community. In this collection, experienced instructors describe their methods for teaching an array of writing and design strategies relevant to content management.

One response arc to this exigence moves through curriculum development. However, in "A Systematic Literature Review of Changes in Roles/Skills in Component Content Management Environments and Implications for Education," Tatiana Batova and Rebekka Andersen (2017) assert that "courses dedicated specifically to component content management have been slow to appear in [technical communication] curricula" (p. 174). I do not find this surprising. Creating new courses requires a trial-and-error period, which could stretch for semesters depending how often a course is offered and how often an instructor is assigned that course. In my department, for example, we offer a new course as a special topics course three times to see how the course does in terms of enrollment before building a master syllabus. The process for introducing a new course is time-consuming in academia, often taking a year or more for all the committee and administrative approvals. Many instructors teach 4/4 loads, and are not provided with professional development time (or compensation) for curriculum work, and therefore may have limited time or means to gain the knowledge needed to rethink curriculum and pedagogy. There simply is not enough time to learn, to adapt, and to redesign curricula and pedagogies, especially when these processes also require that they learn new technologies. But change we must, no matter how comfortable or time-consuming. We must evolve, update, and rethink our curricular frameworks and pedagogical habits as technological changes affect the field.

I want to begin this introduction with a review of three conversational threads connected to content management: the methodology of single-sourcing content, the shift to content-management-driven writing and design, and the craft tradition and its focus on rhetorical value. These threads share a concern for the changing nature of writing practices and their effects on the field and its instruction. I follow that review with a discussion of what I call *networked agency*. Using Etienne Wenger's (1998) notion of building an identity of participation (p. 56), I argue that when writing in what Ann Rockley and Charles Cooper (2012) call "true collaboration," writers can experience both individual and collaborative agencies through a networked system built on the trust gained through the construction of shared content models and standards. I conclude this introduction with summaries of author chapters.

Pedagogical Conversations and Content Management

Technical communication as a discipline has not responded to the challenge of teaching content management with the same dedication or urgency with which it has examined teaching the service course, or web design, or document design, for example. Scholars have tried to engage the community, and in doing so have generated an important conversation about content management and related topics that spans almost three decades. I focus on three threads here. Following single-sourcing, the second thread, running from roughly the mid-1990s to about 2009, focuses on the emerging practices that have been described as a "seismic shift" (a statement first attributed mostly to Dicks, 2010; Hackos, 2000) from the craft model of documentation to single sourcing. The third thread offers a more mature, reflective take on content management, and emphasizes content development, collection, and management, and on publishing strategies and practices (e.g., Boiko, 2005). The last thread seems intent on putting the craftsman model in moratorium. Each thread provides a perspective of technical communication that emphasizes necessary changes to writing practices, and the impact those changes might have on the teaching of writing. I discuss each thread in the next section.

Thread 1: Single Sourcing

Writing practices in technical communication have been drastically changed by a methodology called single sourcing. In *Single Sourcing: Building Modular Content*, Kurt Ament (2002) defines *single sourcing* as a "documentation method that enables" content reuse (p. 1) and repurposing for multiple platforms by breaking it down into manageable chunks (breaking down content into the smallest possible units), which is based on the principle of *granularity* (Ament, 2002; see also Rockley & Cooper, 2012). To compose in chunks, writers work in tandem to create content units that will be assembled later into an information product through a content management system. (Contrast this method with the craft model, where a coherent document is written and formatted by an individual or team as a single-purpose information solution). Among the advantages of the single-sourcing methodology are consistency of branding, standardization of content, efficiency of development and distribution, and, of course, cost savings. Despite single source's promise of efficiency of style and economics, some academics have expressed concern that it could potentially marginalize technical communicators, reducing their role to be little more than a "typist" who merely "assembles writing" (Clark, 2014, p. 71). Sapienza (2007) notes that academics fear that it will "deskill and fragment the profession" (2007, p. 85). The idea of "content reuse," Sapienza says, quoting Lancaster, "does not fit into the academic model, and in many cases contradicts traditional rhetorical principles" (p. 84) because it is "egoless" (Wiess, 2002, p. 3), "context-less" (Eble, 2003), or "*kairos* neutral" (Clark, 2008).

During the mid-1990s to roughly 2009, most of the academic scholarship focused on the introduction of the concept of single sourcing and its impact on the field of technical communication (Rockley, 2001; Williams, 2003; Albers, 2005), on writers and writing (Carter, 2003), and on pedagogy (Eble, 2003; Sapienza, 2007; Robidoux, 2008). Other equally influential publications include discussions about using and teaching XML and the Darwin Information Typing Architecture (DITA), which are covered fully by Becky Jo Gesteland and Jason Swarts respectively in this volume (see also Priestley, Hargis, Carpenter, 2001; Battalio, 2002; Kramer, 2003; Sapienza, 2004; Swarts, 2010; Broberg, 2016; Evia & Priestley, 2016). To build on the concept of single sourcing and the technical communication classroom, I suggest that instructors begin with the 2005 special issue of *Technical Communication* (Volume 50, Number 3) as a means for providing a more extensive background. What I would like to focus on is the pedagogical significance of single sourcing (also referred to as a methodology for writing modular or structured content; see Ament, 2002) and how it affects the writing process and the subsequent impact on pedagogy (see, for example, all articles in the special issue of *Technical Communication*).

In one of those articles, Michelle Eble (2003) notes that the "theoretical principles of single sourcing change the landscape of the technical communication classroom" (p. 345). In fact, she claims that we can no longer teach "only the traditional writing process" (i.e., the craft model of writing), and that we must "introduce students to the process of writing content and information that can be used in various products" (p. 345); that is, to teach single sourcing. We need, she argues, to "redefine the writing process" in order to teach students to "write content that can be reused in various documents" (p. 345). This clearly identifies the exigency we faced during this time period (and still do). Although identifying the problem is vital, we also need to offer technical communication instructors the professional development opportunities that help them learn how to teach single sourcing. This means redefining the rhetorical situation; as Michele Eble (2003) explains, it needs to be "written with all the audiences, purposes, and documents in mind," which depends on the ability of the writer to create "medium-neutral text that can be used in multiple documents" (p. 346), or topic-based writing (which Yvonne Cleary discusses in her chapter in this volume).

During this period, Sapienza (2008) and Charlotte Robidoux (2008) offer the field a pedagogical view of single sourcing. Sapienza (2007) develops a "rhetorical approach to single-sourcing via intertextuality," arguing that single sourcing "is not a set of technologies … but a complex set of rhetorical strategies" that "[involves] planning, linking, arrangement, and deployment of reusable content according to distinct platform constraints and diverse audience needs" (p. 84). This description is important because it reminds us that single sourcing still involves consideration of situation and audience. Building a connection to single sourcing through a hypertextual foundation, Sapienza (2007) argues that "single sourcing is a highly skilled craft that draws on multiple knowledge domains and requires explicit principles consciously applied" (p. 85). Robidoux (2008) also provides a complex view of a single sourcing curriculum based on the "structured writing methods used in many organization" (p. 115). This curriculum, she argues, attends to the "broad theoretical and practical dimensions that can function as an overarching focus for learning" (pp. 120–121). This pedagogy helps us as instructors not only to understand the concept of single sourcing but also to envision a course that relies on it. What works well in Robidoux's proposal of a single-source curriculum is her focus on "structured writing proficiency" and "collaborative teams," both of which are essential for teaching using a single-sourcing methodology. Both pedagogical approaches put us on the road to developing the sound pedagogical approaches Pullman and Gu (2008) identified as crucial for academia's movement into teaching structured writing using single sourcing.

At the heart of this conversational thread is scholarship focused primarily on defining single sourcing as an influence on teaching (e.g., Eble, 2003, Sapienza, 2007; Robidoux, 2008), identifying the challenges it brings to the field (e.g., Williams, 2003), and lamenting its effect on the craft tradition (e.g., Albers, 2003). Little of the scholarship focused on practical pedagogical design. However, the *Technical Communication Quarterly* special issue and the edited collection on content management (Pullman & Gu, 2009) captures the conversation nicely by further defining the problem in terms of pedagogical focus. However, nomenclature was beginning to switch from *single sourcing* to *content management.* In fact, Pullman and Gu (2008), state emphatically that "with all the buzz from the industry about CMSs," there is a "rather *urgent* … need for us to teach content management in our technical communication courses" (p. 3, *emphasis added*). Indeed. This urgency drives the impetus of this collection by providing the professional development instructional need to teach structured writing. Despite the rich discussion of single sourcing at the turn of the century, I believe that for many of us teaching technical communication, even today, single sourcing is still a foreign concept not because instructors refuse to change, but because they lack opportunities for professional development. At this point, at the end of this conversational thread, the nomenclature had already shifted to *content management* as the frame from which we understand ourselves and our discipline.

Thread 2: Content Management

The second conversational thread shifts the focus beyond single-sourcing writing methods to content management and content strategy. One of the first to push the scholarly conversation forward, Bill Hart-Davidson (2010) argues that viewing content management "through the lens of a single concept—single sourcing—has prevented us from seeing," perhaps, the "range of expertise required to ensure that large groups not merely individuals or small teams, write well together" (p. 129). Hence, Hart-Davidson argues for using *content management* rather than *single sourcing* because it "presents a broader, more nuanced view of [content management] and [CMSs] than is typically found in the technical communication literature" (p. 129). For Hart-Davidson, this view is more "nuanced" because it focuses on content management as a "set of practices" rather than a single technique (p. 129). This more nuanced view helps us think beyond simply passing one chunk of content back and forth among a community of writers, editors, information architects, and content strategists. It helps us see the community and the embodied practices that shape its work. As Rockley (2003) so rightly reminds us, "It's about people, not just technology" (p. 350). This sentiment harkens back to Carolyn Miller, who in 1979 made the argument that technical writing (communication) is humanistic and that we should teach it as enculturation, or as an understanding of "how to belong to a community" (p. 617). Teaching it that way enabled us to focus our pedagogies on the rhetorical tradition in addition to its form. Thinking now about the teaching of content management, her argument still holds true, especially when she says that

> to write, to engage in any communication, is to participate in a community; to write well is to understand the conditions of one's own participation—the concepts, values, traditions, and style which permit identification with that community and determine the success or failure of communication.
>
> *(p. 617)*

Members cannot simply participate; they must participate within a shared understanding of a community's practice. This is how members acquire understanding of the conditions of practice.

Content management is a shared practice bound by relationships. The relationships associated with it are important because, as McCarthy, Grabill, Hart-Davidson, and McLeod (2011) describe in their workplace study, "Content Management in the Workplace: Community, Context, and a New Way to Organize Writing," we need to think of content management systems as more than just "technical tool[s]" (p. 369); they are an "essential part of the knowledge-work practices that enable people to manage information in distributed work environments" (p. 369). Similarly, Hart-Davidson (2010) refers to content management as a "set of *practices*" (p. 130, *emphasis added*), as do Andersen and Batova (2015) when they call it an "interdisciplinary practice" (p. 247). Content management, for Andersen and Batova, focuses on the "processes through which production happens" (p. 369) and processes are managed by people. CMSs, Clint Lanier (2012) says, "evolved from the idea that organizations *manage* their information" (*italics in original*, p. 100). I am comfortable with the focus on practices because it shows that content management is a practice shared among members who work in CMSs. Communities form around practices like these, Etienne Wenger (1998) might say, because to participate members must engage in the negotiation of the meaning of a practice, which makes the practice a community-based enterprise. This emphasis suggests more a community than an individual, working collaboratively to manage content with the user experience in mind; it focuses on the community, rather than the artifact.

Thread 3: The Craft Tradition

Part of the initial resistance to embracing the industry shift toward content management involves its contrast to and potential for usurping the traditional Humanities approach to teaching technical communication (see Bridgeford, 2018), referred to most often as the *craft tradition*—the third conversational thread. In his discussion about the technical communication career path, Michael Albers (2003) describes the "craftsman model of production" as one within which "each writer … assigned sections of a [document] … is solely responsible for the content," that is, "with each person handcrafting their own piece" (p. 336). As teachers, this single-writer-single-project perspective simplifies the grading process, aligns with historical models of authorship, and requires little or no technology. The contrasting concepts and strategies between single sourcing and the craft approach to content creation have us perhaps somewhat queasy about teaching single sourcing.

Part of our unsettledness probably centers on how instructors see technical communicators, and thus ourselves. It is likely that we "still view [ourselves] as independent craftspeople" (Andersen, 2014, p. 121), a perspective that differs significantly from the networked writing environment of CMSs. It follows that if "we continue to support and [teach] the craftsman model of technical documentation" (McDaniel & Steward, 2011, p. 195), we are conceivably blocking students' paths to success in this age of content management. But as Filip Sapienza (2007) notes, dismissal of the craft approach is a "source of anxiety" (p. 97) for us in academia, because we are fearful that technical communication is no longer a "one-person operation" (Eble, 2003, p. 336). Embracing that truth would require us to change our ways. As long as we find it difficult to "overcome [that] traditional … approach to document production" (Bacha, 2009, p. 156) or to "break free of the craftsman model" (McDaniel & Steward, 2011, p. 210), we will lag behind important changes affecting technical communicators' writing practices. If the craft approach to writing has truly become "outdated" (Batova, 2018, p. 313), then this discussion thread renders it not only inadequate, but also useless to our pedagogies. I do not believe this is the case, because "craft" always has been reworked to address technological shifts that have affected the field and how we teach writing.

I think it goes without saying that we absolutely want students to enter industry "prepare[d] … to work in [CMS] environments" (Pullman & Gu, 2008, p. 4). However, we do not want to do so at the expense of the craft tradition that has guided us so successfully to this present. But as part of our craft teachings, we also want students to know when, where, and how to question practices. Although I do think that our writing environments have changed with CMSs, I do not believe that craft "has long since passed" (Albers, 2005), and I caution us not to dismiss it too easily. I agree with Mark Baker (2013), who says that "writing remains a craft" (Baker, 2013, p. 201), whether that means writing whole documents or writing modules. As we move from a "document-based perspective to a topic-based," collaborative perspective (Andersen, 2014), I suggest that we resituate craft, that we redefine it in terms of what Rockley and Cooper (2012) call "true collaboration" (p. 224). This notion of true collaboration plays a role in what I am calling *networked agency*, a topic I turn to next.

Networked Agency

From the craft perspective, agency is practiced by individuals who cooperate with each other to complete their work. From a content management perspective, agency is distributed through a digital system, requiring a level of collaboration we have not seen before. For McCarthy et al. (2011) these new kinds of "writing contexts" embody the "values, expectations, and practices shared by a group of writers in a workplace community" (p. 370). Hart-Davidson, Bernhardt, McLeod, Rife, and Grabill (2008) argue for a perspective of content management that sees content management as a "type of conduct," which "permits [them] to explore [content management] as a means to guide decision-making about the creation of knowledge, the arrangement of content, the selection of tools, and the design of work practices associated with the making of text" (p. 10). To emphasize this perspective, for example, an instructor could ask students to create a style guide, an assignment that works well because it situates students in ways that enable them to articulate the purposes of a particular practice such as using one term over another or agreeing to structure content in a specific way.

By pointing to Miller (1989), Hart-Davidson et al. (2007) situate, as she did, a kind of framework based on the "good of the community" (p. 10). Because it is so important to teach technical communication as "how to belong to a community" (Miller, 1979, p. 617), we can embrace the pedagogies that allow us to bridge the approaches associated with the craft tradition and current approaches focused on content management. Swartz (2010), for example, suggests that to make the kind of practical decisions Hart-Davidson (2010) lists, technical communicators must develop a "rhetorical sensibility" (craft) that enables them to "recognize" the arrangement and structure of other texts and to recontextualize them in ways that make sense within a new context of use. These scholars see not a singular CMS, but rather a community of members whose strength is a "networked nature of writing" (McCarthy et al., 2011, p. 368) that aids members in learning to share their work processes and language use as created in a CMS, to create, in fact, a "shared idea of writing" (p. 387). This is how networked agency becomes possible.

A shared idea of writing grows out of the practices of a community. To do so, community members have to participate in ways that enable them to develop what Etienne Wenger (1998) calls an *identity of participation*, his notion of agency, which means that members understand the conditions of their participation, just as Miller (1979) said in her humanistic rationale. This identity of participation helps both newcomers and experienced members participate in meaningful ways and through that participation understand the meaning of a community's practice. In the networked environment of a CMS, a shared idea of writing goes beyond entering text into an electronic field; it means understanding the creation and usage of content models that establish not only the community's ways of working, but also its design for communicating. Technical communicators (whether students or established professionals) need to understand what it means to develop an identity of participation through community-engaged professional efforts. Within a CMS environment, such practices are networked, as is the writing itself. An identity of participation becomes what Wenger might in a CMS think of as *identities of participation*, with all members contributing to the sharing of content, creating a shared identity. When this happens, I think networked agency can happen.

For example, I once asked students to create a new logo and website for my Graduate Certificate in Technical Communication program. While giving their presentation, they explained their decision-making protocol and ways of working as "creative and inventive," explaining that they decided to forego any rules while brainstorming. They kept saying whatever idea came to mind. Later they amassed all the ideas and looked for patterns together. In their evaluative comments, a different student wrote that she did not feel as if she belonged to her group, because her ideas were "being constantly shut down."

Students in the original group demonstrated how identities develop through the negotiation of content models and their application by individual writers within a shared, negotiated space, what Jason Swarts (2010) meant when he referred to "network-building activities," activities where actors engage with each other through the reuse of texts, or single sourcing, and subsequently through the collaborative authoring that engages those texts. What is important to emphasize here is that these texts, what Swarts calls *fractional texts*, "behave like actor networks" within which writers make situated, rhetorical decisions (p. 131). Swartz says that they "mediate and coordinate distributed work" (p. 131); individual writers develop an identity of participation, writing these fractional texts that Swarts argues "become sites in which distributed rhetorical identities and functions coalesce to structure and organize interpretations of the content that they combine to create" (p. 131). When that content becomes combined and networked within a CMS and among all principle actors, it makes possible the creation of identities of participation. Here, multiple writers work "in concert with" multiple content elements in an electronic environment, creating what Rockley and Cooper (2012) call *true collaboration* (p. 224), a condition in which team members work toward a common goal, a "unified content strategy." For this level of collaboration to work, writers must move beyond the traditional, cooperative model they are used to and trust that each other will not only do the work required but also ensure that the entire team has consensus. This level of trust grows out of "building strong [content] models" (Rockley and Cooper, 2012, p. 224) and writing together to determine the various components and structures in ways that embody the organization's enterprise. This capacity for trust is what makes building identities of participation and networked agency possible. Agency happens when writing happens, and writing happens continuously in CMS environments.

Chapter Summaries

The collection moves from defining terms and competencies to teaching the kinds of tasks associated with content management to identifying the kinds of communities in which the practice happens. The authors in this collection provide a variety of pedagogical approaches, along with example assignments, presenting possible avenues instructors can follow as they adjust their pedagogies to address issues associated with content management.

In Chapter 1, George Pullman and Baotong Gu provide a historical timeline of the transition to "systematic content management" that traces the key lexicon of content management terminology, helping us understand what CMSs are and do. They examine the complexities and challenges, which include issues such as the separation of form and content as well as the endangerment of the craft tradition. They conclude their chapter with a description of a content-management-driven pedagogy that can help us design our pedagogies in ways that guide students to think about technical communication from a content management perspective. They conclude with a pedagogical description and example assignment.

Saul Carliner outlines in Chapter 2 what he sees as required knowledge and abilities for this new era of content management and technical communication. Drawing first on the STC (Society for Technical Communication) Competency Model, Carliner describes the nine competencies defined by that organization specific to technical communication. His competency framework includes competencies applicable to interpersonal activities, critical thinking, and decision-making, and to content management (including collaborative writing, dynamic publication, and content strategy). This canopy view helps to introduce instructors to the broad array of the knowledge and abilities required of technical communication students and professionals in contemporary workplaces. Carliner helps us see that content management is broader than one class, and that students can understand and develop the requisite knowledge throughout a curriculum.

In Chapter 3, Liza Potts and Laura Gonzales describe three connected assignments centered on content strategy. They suggest that content strategy involves a vision, action plan, branding, organization, management, and delivery of content. They explain that the purpose of developing a content strategy is to see content management as a "deeply rhetorical practice," and as part of a sound content strategy. As we struggle to "keep pace with industry trends," they argue, we have not done a good job engaging in content management discussions. They describe three interconnected assignments, which they call Landscape/Competitive Analysis, Tone and Style Guide, and the Content Strategy document.

Yvonne Cleary, in Chapter 4, describes her approach to teaching topic-based writing in technical communication. She outlines the evolution of topic-based writing, a timeline she traces to Robert Horn's information mapping. As part of her teaching outline, she reviews the seven basic characteristics of Mark Baker's *Every Page is Page One: Topic-based Writing in Technical Communication*. She describes an assignment for teaching topic-based writing that she designed to be flexible because students use open-source software. She concludes with a comment on the adaptability of the assignment for collaborative work, expansion of that assignment unit by developing it into a full course on structured writing, and preparedness of instructors for the ever-changing field of technical communication.

In Chapter 5, Bill Williamson and Scott Kowalewski discuss the value of teaching usability studies as part of user-centered design. Content management and single-source systems, they argue, provide designers with more meaningful tools for engagement with this methodology in terms of pedagogy. The authors describe their approaches to teaching usability studies assignments in courses for both majors and nonmajors, describing activities they use for thinking about the application of this strategy in CMSs. They conclude with a list of resources to help instructors get started on implementing usability studies into their classes.

Becky Jo Gesteland (McShane) argues in Chapter 6 that it is important to teach students how to code in XML. Extending one of her earlier arguments (McShane, 2009), she suggests a sequence of topics and assignments connected to various readings that help students understand content management and the need for learning to code. She emphasizes four steps in the process, which include deciding on a topic, evaluating appropriate tools, creating an information model, and developing the content.

With Chapter 7, Elise Verzosa Hurley and Amy C. Kimme Hea describe the functions and affordances of social media as CMSs. They argue that to succeed, organizations should be "preemptively reaching audiences" through the dynamic content in social media systems such as Twitter, Facebook, and YouTube. Helping students develop a "rhetorical understanding" of reach is vital because technical communicators are poised to step into social media positions. The three assignments they describe—a social media analysis, a social media technology quick reference card, and a social media campaign—work together enabling students to gain the skills needed for working in distributed environments.

In Chapter 8, Carleigh Davis and Michelle F. Eble take a case-based, human-centric approach to audience analysis that involves determining the wants, needs, and motivations of audience members. Although in practice this parallels how we have always taught technical communication, the authors recast this process with CMSs in mind. Their approach makes use of a six-step heuristic that provides instructors a framework for helping students plan, write, and deliver audience-focused content suitable for a CMS. In doing so, they draw on the concept of *phronesis* in ways that guide decision-making for constructing knowledge, arranging content, selecting tools, and establishing work practices. They conclude with two ready-to-use scenarios.

In Chapter 9, Jason Swarts describes a pedagogy for teaching the DITA. His approach helps teachers help students understand the rhetorical challenges of writing and repurposing structured content. To that end, he provides various lessons for teaching bilocational and translocational meaning, which signifies content that can be stored in more than one place and incorporated into information products in multiple ways. He provides examples of DITA code and information mapping with effective assignments for teaching structured writing.

In Chapter 10, William Hart-Davidson and Ben Lauren describe a process they call *writing stewardship*, which means using one's expertise in writing to help others in an organization work well. This concept is important for instructors of technical communication because as they move beyond the entry level, they need to understand how they better manage projects and develop skills in negotiation and facilitation that embody the role of writing stewardship.

Kirk St.Amant, in Chapter 11, describes how cultural factors affect content expectations, and argues that because content is multifaceted, it affects all aspects of human life. He emphasizes that the cultural-rhetorical factors of various genres depend on the purpose and audience of a particular cultural, especially given that cultures carry with them different expectations. His approach to teaching cross-cultural communication focuses on localization and translocation, which involve revising content to meet the expectations of a particular culture. He provides example assignments that grow out of that practice.

In the Afterword, Carlos Evia and Rebekka Andersen take us through the stages of Content 4.0 as imagined by Joe Gollner and Marie Girard. This model illustrates the way we assemble, transform, and render dynamic content. In true Afterword fashion, they move us beyond thinking of managed content solely as static objects, and help us to see the forces that shape that content. They discuss four forces, including the need to publish to many channels, the need to provide seamless content experiences, the need to redefine our relationship to subject matter experts, and the need for continuous content development. They conclude with suggestions for how faculty can gain the requisite skills and knowledge to advance content management curricula.

References

Albers, M. (2003). Single sourcing and the technical communication career path. *Technical Communication, 50*(3), 335–343.

Albers, M.J. (2005). The future of technical communication: Introduction to this special issue. *Technical Communication, 52*(3), 267–272.

Ament, K. (2002). *Single sourcing: Building modular documentation.* Norwich, NY: William Andrew Publishing.

Andersen, R., & Batova, T. (2015). The current state of component content management: An integrative literature review. *IEEE Transactions on Professional Communication, 58*(3), 247–270. doi:10.1109/TPC20162516619

Andersen, R. (2014). Rhetorical work in the age of content management: Implications for the field of technical communication. *Journal of Business and Technical Communication, 28*(2), 115–157. doi:10.1177/1050651913513904

Bacha, J. (2009). Single sourcing and the return to positivism: The threat of plain-style, arhetorical technical communication practices. In G. Pullman & B. Gu Eds., *Content management: Bridging the gap between theory and practice* (pp. 143–160). Amityville, NY: Baywood Publishing Company, Inc. New York: Taylor & Francis.

Baker, M. (2013). *Every page is page one: Topic-based writing for technical communication and the web.* Laguna Hills, CA: XML Press.

Batova, T. (2018). Work motivation in the rhetoric of component content management. *Journal of Business and Technical Communication, 32*(3), 308–346. doi:10.1177/1050651918762030

Batova, T., & Andersen, R. (2017). A systematic literature review of changes in roles/skills in component content management environments and implications for education. *Technical Communication Quarterly, 26*(2), 173–200. doi:10.1080/10572252.2017.1287958

Battalio, J.T. (2002). Extensible markup language: How might it alter the software documentation process and the role of the technical communicator? *Journal of Technical Writing and Communication, 32*(3), 209–244.

Boiko, Bob. (2005). *Content management bible*, 2nd ed. Indianapolis, IN: Wiley Publishing, Inc.

Bridgeford, T. (2018). *Teaching professional and technical communication.* Logan, UT: Utah State University Press and Boulder, CO: Colorado State Press.

Broberg, M.S.E. (2016). A decade of XML—And a new procurement and lessons learned. *Technical Communication, 63*(1), 6–22.

Carter, L. (2003). The implications of single sourcing for writers and writing. *Technical Communication, 50*(3), 317–320.

Clark, D. (2008). Content management and the separation of presentation from form. *Technical Communication Quarterly, 17*(1), 35–60. doi:10.1080/10572250701588624

Clark, D. (2014). Rhetorical challenges and concerns in enterprise content management. In J. Vom Brocke & A. Simons (Eds.), *Enterprise content management in information systems research* (pp. 63–74). New York: Springer.

Dicks, R.S. (2010). The effects of digital literacy on the nature of technical communication work. In R. Spilka (Ed.), *Digital literacy for technical communication: 21st century theory and practice* (pp. 51–82). New York, NY: Routledge Publishing.

Eble, M. (2003). Content vs. product: The effects of single sourcing on the teaching of technical communication. *Technical Communication, 50*(3), 344–349.

Evia, C., & Priestley, M. (2016). Structured authoring without XML: Evaluating lightweight DITA for technical documentation. *Technical Communication, 63*(1), 23–37.

Gesteland McShane, Becky Jo. (2009). Why we should teach XML: An argument for technical acuity. In George Pullman and Baotong Gu (Eds.), *Content management: Bridging the gap between theory and practice* (pp. 73–85). Amityville, NY: Baywood Publishing Company, Inc.

Hackos, J. (2000, January). Trends for 2000s: Moving beyond the cottage. *Intercom*, 6–10.

Hart-Davidson, W. (2010). Content management: Beyond single sourcing. In R. Spilka (Ed.), *Digital literacy for technical communication: 21st century theory and practice* (pp. 128–143). New York: Routledge.

Hart-Davidson, W., Bernhardt, G., McLeod, M., Rife, M., & Grabill, J.T. (2007). Coming to content management: Inventing infrastructure for organizational knowledge work. *Technical Communication Quarterly*, *17*(1), 10–34. doi:10.1080/10572250701588608

Hart-Davidson, W., Bernhardt, G., McLeod, M., Rife, M., & Grabill, J.T. (2008). Coming to content management: Inventing infrastructure for organizational knowledge work. *Technical Communication Quarterly*, *17*(1), 10–34. doi:10.1080/10572250701588608

Kramer, R. (2003). Single source in practice: IBM's SGML toolset and the writer as technologist, problem solver, and editor. *Technical Communication*, *50*(3), 328–334.

Lanier, C. (2012). Accounting for the human element when planning for a content management system. *Technical Communication*, *59*(2), 99–111.

McCarthy, J., Grabill, J.T., Hart-Davidson, W., & McLeod, M. (2011). Content management in the workplace: Community, context, and a new way to organize writing. *Journal of Business and Technical Communication*, *25*(4), 367–395.

McDaniel, R., & Steward, S. (2011). Technical communication pedagogy and the broadband divide: Academic and industrial perspectives. In A.P. Lamberti & A.R. Richards Eds., *Complex worlds: Digital culture, rhetoric, and professional communication* (pp. 195–212). Amityville, NY: Baywood Publishing Company, Inc. New York: Taylor & Francis.

Miller, C. (1979). A humanistic rationale for technical writing. *College English*, *40*(6), 610–617.

Miller, C. (1989). What's practical about practical writing? In B. Fearing & W. K. Sparrow (Eds.), *Technical writing: Theory and Practice* (pp. 14–24). New York: Modern Language Association.

Priestley, M., Haargis, G., & Carpenter, S. (2001). DITA: An XML-based technical documentation authoring and publishing architecture. *Technical Communication*, *49*(3), 352–367.

Pullman, G., & Gu, B. (2008). Guest editors' introduction: Rationalizing and rhetoricizing content management. *Technical Communication Quarterly*, *17*(1), 1–9. doi:10.1080/10572250701588558

Pullman, G., & Gu, B. (2009). *Content management: Bridging the gap between theory and practice.* Amityville, NY: Baywood Publishing Company New York: Taylor & Francis.

Robidoux, C. (2008). Rhetorically structured content: Developing a collaborative single-sourcing curriculum. *Technical Communication Quarterly*, *17*(1), 110–135. doi:10.1080/10572250701595652

Rockley, A. (2001). The impact of single sourcing and technology. *Technical Communication*, *48*(2), 189–193.

Rockley, A. (2003). Single sourcing: It's about people, not just technology. *Technical Communication*, *50*(3), 350–354.

Rockley, A., & Cooper, C. (2012). *Managing enterprise content: A unified content strategy (2nd ed.).* Indianapolis, IN: New Riders.

Sapienza, F. (2004). Usability, structed content and single sourcing with XML. *Technical Communication, 51*(3), 399–408.

Sapienza, Filipp. (2007) A rhetorical approach to single-sourcing via intertextuality, *Technical Communication Quarterly, 16*(1), 83–101.

Sapienza, F. (2008). A rhetorical approach to single-sourcing via intertextuality. *Technical Communication Quarterly, 16*(1), 83–101. doi:10.1080/10572250709336578

Swarts, J. (2010). Recycled writing. Assembling actor networks from reusable content. *Journal of Business and Technical Communication, 24*(2), 127–163. doi:10.1177/1050651909353307

Wenger, E. (1998). *Communities of practice: Learning, meaning, and identity.* New York: Cambridge University Press.

Williams, J. (2003). The implications of single sourcing for technical communicators. *Technical Communication, 50*(3), 321–327.

PART I

Definitions

1

RECONCEPTUALIZING TECHNICAL COMMUNICATION PEDAGOGY IN THE CONTEXT OF CONTENT MANAGEMENT

George Pullman and Baotong Gu

GEORGIA STATE UNIVERSITY

Chapter Takeaways

- Content management (CM) is not about software.
- CM is what writing has become.
- Separation of form and content is essential to writing now.
- Single sourcing is key to consistency of message.
- Content managers need to craft messages that can exist in different formats.

Content management has been a buzzword in the field of technical communication for about a decade, but more so in the last few years. If you have been in this field for even a couple of years, you probably could not have helped noticing the term popping up around you: during your online search for technical communication information, your library research for an article you are working on, your browsing session on the latest publications in our field, the annual Association of Teachers of Technical Writing or Society for Technical Communication conferences, or sometimes even class discussions on the latest developments in technical communication. The truth is students have already heard the term, and some have even designed or customized their own CM software; your colleague next door is adopting a CM-driven pedagogy. Granted, there is still a wide spectrum of knowledge and skills when it comes to CM: from a cursory recognition of the term to full ability to customize or even design one's own system, it is becoming more relevant to what we do as technical communication instructors. You ask yourself: What exactly is CM? What does that have to do with me? Can I afford not knowing what it is? What are some of the key terms? What are the fundamentals I need to know about CM? What does a CM-driven pedagogy look like? In this chapter, we attempt to answer some of these questions and explicate on why a CM approach is essential to today's technical communication, in what way it is challenging our traditional conceptualizations of rhetorical design and revolutionizing our field, and how it can inform our pedagogical practice both theoretically and operationally.

Over the last decade or so, the field of technical communication has witnessed nothing short of a paradigm shift: from the conceptualization of writing as a rhetorical act of composing words on paper to that of content creation and asset management. Content creation and management is about composing with words, images, video, data, metadata, and the containers that deliver the content. While documents are inert and autonomous, digital content is dynamic and context responsive. For us, CM is a systematic way of thinking about (and executing) how to create, share, archive, and track the efficacy of information; who wrote which version(s); which version is current; how old it is; its siblings and descendants; its planned lifecycle; and so on. We are talking about designing and creating content for an information ecosystem and the complex act of asset management. Thus, in the digital era, technical writers do not write so much as they create, curate, and manage content. They still have to compose efficient prose, of course, but they need to have a much broader skill set as well as a deeper understanding of information and workflow; not just technical knowledge of the field they write to the world about, but knowledge of such things as content strategy, communication medium, software applications, CM systems (CMSs), and so on.

Jonah Winters's description of his work with the Bahá'í Library Online is a perfect illustration of the contrast between the traditional method of managing content in the pre-CMS era and the new method of CM.

- **The old way** (a partial summary):
  1. Documents are emailed to him in a variety of formats.
  2. He manually converts all the formatting to HTML.
  3. He uses a number of professional, often complex programs to create content: *Photoshop* to edit images, *RTFtoHTML* to convert *Word* documents to HTML, *Dreamweaver* to edit HTML, and numerous helper apps.
  4. He compiles all document information he can find and creates a blurb for index pages.
  5. He adds links to the new site to the index pages and possibly two to five other cross-reference links.
  6. He adds the document to his *Word*-based list catalog.
  7. *Making corrections:* he manually makes any corrections in the HTML document, tests, and uploads the new document.
  8. If anything more than cursory changes are made, he re-converts the entire document.
  9. He searches for all links that need to be updated and changes them one by one.

His time commitment: two hours per day every day, every month, every year.
Pros: full control over formatting and cataloguing.
Cons: only works well with sites of a few hundred files (the Bahá'í Library now has 18,000 files); site ends up being full of errors, dead links, outdated content, and limited cross-referencing, and becomes wholly unmanageable for an individual or even a small team of individuals.

- **The new way** (a complete summary):

  1. He programs an interface in PHP and sets up a back-end database.
  2. All further content is added by public users, not by himself.
  3. Users log in, then are taken to an upload/editing screen, where they contribute new content or update preexisting content.
  4. When users click "Update," the system (1) formats basic HTML automatically, (2) prepares the blurb and as many index entries as are cross-referenced, (3) catalogs the document, and (4) displays the document to the public.
  5. *Making corrections:* users log in and make any updates they wish; the system then stores and catalogs the corrections across the entire site.

  His time commitment: 1,000 hours up front (300 initial programming, 200 upgrades and module additions, and 500 hours for manual data migration, copying all information for each file into the database in each appropriate field), less than 30 minutes per day from then on.
  Pros: anyone can register, upload content, correct errors, and add cross-references; site is easy to navigate and search; content is always up to date; site can and soon will house millions of files; site can handle documents in any language.
  Cons: limited flexibility; significant up-front time requirements.

(Adapted from Winters, 2017)

The old-way scenario is representative of how most companies used to manage their content, especially during the static web days. This transition to systematic CM is thus redefining the role of technical communicator from that of a writer to that of a content manager. It is also placing new demands on today's technical communicators as it requires a new skill set, a new approach to technical communication, and a new big-picture mindset about managing content. As a result, it is also placing new demands on us technical communication instructors as it dictates new pedagogical approaches that can accommodate the changing needs of today's technical communicators.

A Brief History of Content Management Systems

CMSs, software applications for systematic management of content/information, began to emerge in the early 1990s. According to Jonah Winters, a rough timeline runs like the following:

First CMSs (1992–1995)

The first CMSs emerged in the early 1990s. Most of these early CMSs were cumbersome and expensive, including, for example, RAINMAN (Remote Automated Information Network Manager), developed by AOL in 1992 (Levitt, 2011).

Emergent Period for Open-source CMSs (1995–1999)

The first open-source CMS, named Wiki Wiki, was designed and used for Portland Pattern Repository in 1995, which was also the year PHP was created. In 1997, PHP was retooled for much better versatility, stability, and ease of use. Although a number of PHP-based CMSs were born in this period, most CMSs in use were those proprietary ones used commercially. This is also the period when three important open-source tools came into being: the database program MySQL, the operating system Linux, and the web server software Apache (Winters, 2017).

Booming Period of Open-source CMS (2000–2005)

The creation of the four critical tools of open-source software—Linux, Apache, MySQL, and PHP—paved the way for an "internet revolution" that saw the exponential increase of open-source websites from thousands to millions as well as of open-source CMSs. This period is when some well-known open-source CMSs came into being, such as Mambo and Drupal (Winters, 2017). WordPress arrived in 2003. It is also the period when a large number of CMSs went through "a massive wave of mergers and acquisitions," leaving many users high and dry without support (Kanga Internet, n.d.).

Period of User Orientation and Customization (2006–present)

Many of the CMSs created in the previous periods afforded little flexibility for customization to accommodate users' needs. The CMS industry was lagging seriously behind the evolving needs of users. It was not until "somewhere around 2009 [that designers] caught up with [users] on the management side" (Tolvanen, 2013). "It wasn't long before the usability, affordability, and ideology driving the popularity of open source began to pose a major threat to the commercial software industry" (Whitehead, 2014). As of October 26, 2017, WordPress, one of the most popular open-source CMSs, holds a 28.9% market share of web CMSs (W3Techs, 2017).

Defining Key Content Management Related Terms

Various terms related to CM have surfaced. To someone new to this field, understanding the meanings of these terms and the distinctions among them is essential. For the sake of our discussion, we briefly define the following terms:

- Content.
- Content management.
- Content management systems.
- Enterprise content management.
- Component content management.
- Content strategy.

There are more than nuanced distinctions among these terms, and we define them and outline the main differences.

Content

Content is seldom defined in books and articles on CM. We tend to assume that we all understand and agree on what *content* means when in fact we do not. The fact that Bob Boiko (2005) devotes the first five chapters of his seminal work *Content Management Bible* to defining the term "content" indicates that we indeed have very different conceptualizations of the term. Various alternatives to *content* have also been suggested, adding to the potential confusion: *content, data, information, knowledge, asset* ... For our purpose here, we limit our discussion to the differences between *content* and *data* (or *raw data* to be exact). *Data*, in our conceptualization, means raw information not processed for any specific purposes. According to Deane Barker (2016), there are two key differences between *content* and *raw data*: (1) content is *created* differently, and (2) content is *used* differently. He therefore defines *content* as "information produced through [the] editorial process and ultimately intended for human consumption via publication." Although this definition captures the essence of CM, the *word* "publication" is somewhat ambiguous and could mean the act of publishing content in some form or delivering it to the end user. To simplify the matter, we define *content* as data that has been edited to accommodate specific user needs.

Content Management

CM "is a set of processes and technologies that supports the collection, managing, and publishing of information in any form or medium" (Wikipedia, n.d.). As a result of its more popular sister term "content management system," "we tend to look at content management as a digital concept, but it's been around for as long as content. For as long as humans have been creating content, we've been searching for solutions to manage it" (Barker, 2016). As "the process for collection, delivery, retrieval, governance and overall management of information in any format" (Kiwak, n.d.), CM may not necessarily involve digital technologies, although in today's environment it is hard to imagine it not doing so. Another term that has often been used in its place is "asset management," although this is more context dependent because it could denote other types of assets, such as financial assets. Whichever the term, *content management* describes the process of content lifecycle from its creation to its presentation, delivery, storage, reuse, etc.

Content Management System

A CMS is "a software application or set of related programs used to create and manage digital content" (Churchville). Put in a different way, "A content management system (CMS) is a software package that provides some level of automation for the tasks required to effectively manage content" (Barker, 2016). These definitions imply two key aspects to CMSs: (1) the content is digital; (2) it involves some form of digital technology. This then begs the question: did CMSs exist in pre-digital era? Many will argue they did not, and there are good reasons for such an argument: if CMSs involve digital technology, how could they have existed before the digital age? In our opinion, though the term "content management system" is a product of the digital age, the concept is not necessarily new, and the term "system" does not necessarily have to refer to digital systems. Your physical filing system in the office where you manage all teaching materials, student information, assignment files, and so on, is a primitive form of CM; on a very small scale, of course. In a similar way, the library's pre-digital-era cataloging system is also a CMS, but on a much larger and complex scale. Of course, the term "content management system" was indeed coined in the digital age and is used exclusively to denote digital systems. However, what we are arguing here is that the concept and the practice of managing our information systematically are not new creations.

Enterprise Content Management (ECM)

It is not enough just to have a CMS managing your content. Your content must serve the business purposes of your enterprise, so you must have the right content in the right structure and the right CMS to manage the content in the right way. According to the Association of Information and Image

Management (AIIM, n.d.), "effectiveness, efficiency, compliance, and continuity all combine, in different proportions, to drive the business case for content management in most organizations." ECM emerged to address such a need. ECM, as defined by AIIM, refers to

> the systematic collection and organization of information that is to be used by a designated audience ... a dynamic combination of strategies, methods, and tools used to capture, manage, store, preserve, and deliver information supporting key organizational processes through its entire lifecycle.

ECM is not a system but more a concept informed by a particular approach. "ECM has evolved into a term more associated with business process and content strategies than as an enterprise content management system" (OnBase, n.d.). It is "a system solution designed to manage an organization's documents. Unstructured information—including Word documents, Excel spreadsheets, PDFs and scanned images—are stored and made accessible to the right people at the right time" (Laserfiche, n.d.). According to AIIM, there are five key elements of ECM:

1. *Capture:* entering content into the system.
2. *Manage:* what you do to the content, so it can be found and used by whomever it is intended for.
3. *Storing:* finding it an appropriate home in your infrastructure, whether it is a formal CMS or other information solution.
4. *Preserve:* long-term care—archiving, if you will; the practice of protecting it so it can be utilized however far into the future the organization needs it to be available.
5. *Deliver:* putting the information in the right people's hands right when they need it to be there.

Given such descriptions and definitions, it is clear that ECM "is an umbrella term for the technology, strategy and method used to capture, manage, access, integrate, measure and store information" (OnBase, n.d.). In this sense, it is similar to the concept of CM in that it refers to a set of processes, but different in that it serves specific business purposes. It is also, in a sense, similar to another concept: content strategy, in that they both imply a deliberate approach to managing content. We will discuss content strategy in detail later in this chapter.

Component Content Management (CCM)

Component content management is a relatively new category in this CM-related glossary. As the term "component" indicates, in CCM content and data are first segmented/chunked into "components." The size of the components varies, depending on well-defined structural needs. They can be as small as a word or as big as a few paragraphs. Each component is then properly labeled, typically in XML (Extensible Markup Language) format, for future use and reference. When the need arises to create an appropriate "document," different components are then selected and retrieved from the database and edited into usable content. This repurposing and reuse of the components (i.e., datatized content) can be a recurring process to minimize the work and to avoid repetition of content creation. The process of content segmentation is sometimes referred to as "datatizing"—the process of segmenting content into pieces of reusable data to be stored in databases. As Trotter (2008) puts it, "rather than storing documents, they store and manage the small re-useable components that are used to assemble documents." *Enterprise content management*, therefore, refers to the process of "manag-[ing] content at a granular level (component) rather than at the document level. Each component represents a single topic, concept or asset (for example an image, table, product description, a procedure)" (Wikipedia, n.d.).

Three factors have driven the development of CCM (Gilbane, n.d.):

1. Single sourcing: original content for all uses is created in a single-source format and could easily generate different formats in the end product, such as print, HTML, etc.
2. Reuse: components created in single sourcing can then be repurposed and reused.
3. Translation and localization: chunked content lends itself to more efficient translation. Instead of sending whole documents out for translation, the translation team can focus just on those components that will be used.

The two key aspects that set CCM apart from traditional CM are: (1) while traditional CM manages content at the document level, CCM manages it at the granular level, thus rendering the content a much greater level of flexibility and adaptability; (2) content can be repurposed and reused, thus reducing the amount of time and cost required for content production.

Content Strategy

Content strategy, the cornerstone of CM, is receiving increasing emphasis in how it directly correlates to the success of CM. Content strategy entered the picture of the CM landscape relatively recently as people began to realize simply having a big-budget CMS in place does not guarantee the success of an organization's CM endeavor. As we struggle to grasp what content strategy really entails, various definitions have emerged, sometimes with substantive conceptual differences.

Kristina Halvorson, author of *Content Strategy for the Web*, offers what is considered the de facto definition of content strategy: "planning for the creation, delivery, and governance of useful, usable content" (quoted by Bussolati, n.d.). Several authors offer definitions along a similar line (Bussolati):

- Planning for the creation, aggregation, delivery, and useful governance of useful, usable, and appropriate content in an experience. (Margot Bloomstein, Principal of Appropriate, Inc.)
- Content strategy encompasses the discovery, ideation, implementation and maintenance of all types of digital content—links, tags, metadata, video, whatever. (Robert Stribley, Information Architect at Razorfish)
- A plan for adding unique, expert, and indexable content to your site on a regular basis. (Mark O'Brien, author of *A Website that Works: How Marketing Agencies Can Create Business Generating Websites*)

Although these definitions more or less capture the essence of content strategy, there is one point of divergence that others have issues with: the implementation aspect. Rahel Bailie, coauthor of *Content Strategy: Connecting the Dots between Business, Brand, and Benefits*, contends:

> Content strategy deals with the planning aspects of managing content throughout its lifecycle, and includes aligning content to business goals, analysis, and modeling, and influences the development, production, presentation, evaluation, measurement, and sunsetting of content, including governance. What content strategy is not is the implementation side. The actual content development, management, and delivery is the tactical outcomes of the strategy that need to be carried out for the strategy to be effective.
>
> *(quoted by Bussolati, n.d.)*

Bussolati agrees with Bailie: "I think Halvorson's de facto definition still is relevant … I would add: content strategy is what guides content teams to best processes for each stage of the content cycle." One point is clear from Bailie's contention: content strategy involves the planning side of things, not the actual implementation. Aside from this point of divergence, all agree on the importance of content strategy to an organization's CM process. We think Morten Rand-Hendriksen's definition (2016) best captures the magnitude of this significance: "content strategy is the art of bringing the right content to the right person in the right place at the right time and in the right context." This argument is echoed by Hannah Smith (2014), who conceptualizes content strategy as a "vision": "content strategy is the high-level vision that guides future content development against a specific business objective." Smith argues that "content creation without strategy often leads to disparate content with no core themes or purpose."

Good CM practice, therefore, starts with a good content strategy, just as good teaching must be guided by a good pedagogical approach. Similarly, the teaching of CM in our technical communication classrooms must start with the teaching of content strategy, which we will elaborate on in the later sections.

Hosted vs. Nonhosted CMSs

With a plethora of CMSs out there, one thing a CMS beginner needs to understand in selecting the appropriate CMS is the distinction between hosted and nonhosted systems. A hosted CMS is one where the provider hosts your website or blog for you. A hosted CMS provides predefined themes and built-in server functionality so that you do not have to worry about working with databases and servers. A *nonhosted* CMS, also called a *self-hosted* CMS, on the other hand, refers to "a situation where you (the client) do all of the hosting yourself on your own server, as opposed to having a third party handle it for you on theirs" (Crouch, 2017).

Take the popular WordPress CMS, for example. There are both hosted and nonhosted versions available. WordPress.com is the hosted version while WordPress.org is the nonhosted version. Table 1.1 outlines the differences between the hosted and nonhosted versions:

These pros and cons pretty much sum up the differences between most hosted and nonhosted CMSs. Leading hosted CMSs include WordPress.com, SquareSpace, Weebly, Wix, Web.com, Business Catalyst, LightCMS, HiFi, and Webpop, while major self-hosted CMSs include WordPress.org, Drupal, Joomla, and Expression Engine (Bollinger, 2014).

So, what CMS should you choose? Before you make this decision, you need to ask yourself a few important questions: how much coding

TABLE 1.1 Differences between Hosted and Nonhosted Versions

	Pros	*Cons*
WordPress.com	• Simple to set up • Support available • Always up to date • Managed server • Ability to connect with other blogs	• Limited themes and plugins • No code-level access • No uploading themes and plugins
WordPress.org	• Massive theme and plugin library • Code-level access • Option to customize everything • Custom designed sites	• Need to manage your own server • Need to install and upgrade platform • Plugin conflicts • Forum support only

(*Adapted from 99Robots, 2014*)

knowledge do you have? How much time and money do you have to invest in your web development? How complex will your site be? How much customization will you need? What are your priorities: ease of use, cost effectiveness, or customizability? If you prefer ease of use, availability of technical support, reliability of the system, constant upgrades, and good value, then a hosted CMS is probably right for you; on the other hand, if your main priorities are customization, the option to incorporate custom properties, back-end system integration, then a self-hosted CMS is probably your right choice (We Make Websites, 2014).

Processes, Challenges, and Complexities

A good understanding of the processes, challenges, and complexities of CM is a prerequisite for any discussion on CM. After all, CM represents a radical departure from our traditional technical communication processes, and it involves a high level of complexities new to technical communication instructors.

Content Management Process

If you conduct a survey on the CM processes of 100 organizations, you are likely to get 120 different processes. No two organizations share the same CM process simply because each organization has different business goals, different philosophies, different CMSs, different content creators and editors, and different everything from the next organization. Likewise, authors on CM have identified differing processes. Bob Boiko (2005) identifies three phases of CM: collect, manage, and publish; Gerry McGovern also sees three: creation, editing, and publishing; both JoAnn Hackos and Ann Rockley identify four components: authoring, repository, assembly, delivery/publishing (CMS Review, n.d.).

What is not overtly apparent from these processes but utterly important in our opinion, however, is planning, which is where content strategy comes in. The CM workflow identified in "Seven Stages of the Content Lifecycle" (CMS Review, n.d.) does seem to recognize the planning aspect as it identifies a seven-step process workflow that includes organization, workflow, creation, repository, versioning, publishing, and archives. In this sense, Rick Allen's (2012) article identifies eight critical elements and tasks in designing your CM workflow, which we think is rather illuminating:

1. *Start with content strategy:* define your content goals, content need, role assignment for planning, creation, and governance of content.
2. *Interview content stakeholders:* meet with your content creators, editors, reviewers, approvers, and anyone who interacts with content.

3. *Define CMS roles:* define the roles of everyone involved in the process, including requesters, creators, editors, approvers (owners), publishers, and so on.
4. *Conduct a CMS task audit:* identify all the tasks involved in the web publishing process.
5. *Map out your CMS task:* put all your tasks into context and determine the interrelationship between tasks.
6. *Integrate your content planning tools into your CMS workflow:* determine your style guides, templates, checklists, etc., to ensure consistency.
7. *Test CMS workflow:* have your stakeholders test out the workflow with actual tasks, both common and uncommon, to identify any potential issues.
8. *Gain support for CMS workflow:* involve your stakeholders early in the process for better chances of gaining their support.

Planning here in Allen's conceptualization of the CM workflow is duly given utmost importance as the process starts with content strategy, a critical step that determines the successful implementation of a sensible CM process.

Complexities and Challenges

For anyone new to CM, the complexities of CM may seem overwhelming, even daunting. The traditional rhetorical approach of single-document design has suddenly morphed into a complex task set involving a whole array of stakeholders, a multitude of tasks, a variety of skill sets, not to mention a technology system typically out of the cognitive reach of most technical communicators. It seems in CM the technical communicator suddenly has to assume a multiplicity of roles: the content strategist, the content creator and editor, database manager, web designer, technical expert, content marketer … Granted, you will not have to assume all these roles and a typical CM process will involve a team of players, but a good understanding of these roles and how they interplay is essential.

Despite these complexities, given time, skills can be acquired, roles familiarized, technology learned. What poses greater challenges is the mindset: how we can transition from the traditional rhetorician/writer to that of a content manager. There are two prominent issues associated with this transition to CM: (1) the separation of form/presentation and content, and (2) the preservation of the craft tradition.

Separation of Form and Content

As Dave Clark has noted,

> on its face the call for a fundamental, technological separation of presentation from content is in tension with our historical interest in creating

> documents that unify content and presentation, and increasingly we are hard coding that tension into CMS and Web CMS.
>
> *(2008, p. 41)*

Rhetorical practice has a long history of advocating the unity of form and content. We have been long taught that content and format must be designed simultaneously to make documents effective and that form, if not unified with content, is meaningless. This separation of form and content, a central feature of CM, especially CCM, is then at odds with the conventional wisdom of the rhetorical teachings.

At the same time, this separation of form and content is foundational to CM, and there are compelling reasons for doing so:

> The advantages of a separation, then, can be compelling. In content management, whether we are discussing website content management, small, in-house single sourcing systems, or enterprise-wide content management (ECM), separating content from presentation can not only save time but allow for rapid reuse and repurposing of content. A single piece of content, properly marked and stored, can automatically and simultaneously appear in user manuals, help files, and press releases that can in turn be automatically altered to appear in print, on the Web, or on mobile devices. Once initial designs are created, fonts, colors, and layout are added on the fly for the specifics of each genre and/or medium, and with, for example, a simple change to a style sheet, aesthetic changes can easily be applied to past as well as future documents, making it easy to maintain organizational consistency.
>
> *(Clark, 2008, p. 36)*

We argue, however, that this separation of form and content is deceptive and at best an early, transient phase in the lifecycle of CM. This seeming separation happens when content is segmented and datatized for storage purposes. When these segmented content is eventually retrieved from the database and edited and molded into its end product—say, end-user documentation—the content is eventually reunited with form. As Clark has argued, "content and presentation are never separated, because even the most poorly formatted Notepad document has a presentation: layout, fonts, paragraph breaks, capitalization, headings. Authors writing for content management work with interfaces that offer authoring-specific presentations of their content" (p. 44). Clark suggests thinking of this "separation" in two distinct ways:

- **Content is complete texts. Presentation is output structure, navigation, and visual style.** In this formulation, most common in Web design and in more rudimentary CMS and single sourcing, presentation is a content wrapper that includes navigation (e.g., links, table of contents, indices), output layout, and the visual style for the content. Content, the authored text and images of a genre (e.g., blog entry, press release, home page), is stored as a unit. Because it is stored as a unit, the text has a predefined structural presentation (headings, paragraphing), and visual style (font choices, color, etc.) is linked to content via that structure.

- **Content is content modules. Presentation is output structure, navigation, visual style, and genre definition.** Rather than being stored as complete documents, content is broken into component parts, at a level of granularity defined by information models (e.g., at the section, paragraph, sentence, or even word level). Content is then divorced from a document-centric structure and can be assembled in myriad ways. Presentation, then, includes the process of building custom documents, creating genres of documents on the fly based on organizational definitions. (p. 45)

Clark's conceptualization is significant and helpful in demystifying this prominent issue. This reconciliation between the separation of form and content in CM on the one hand and the rhetorical principle of the unity between form and content on the other is important, especially to technical communicators new to the CM arena. This is why we argued earlier that a good understanding of the process of CM is essential.

Preservation of the Craft Tradition

The CM approach gives rise to another prominent issue in the field: the endangerment of the craft tradition, addressed by Bridgeford in the Introduction of this collection. The craft tradition, as summarized by Bridgeford, refers to the argument that writing is a craft that depends on the autonomy of the writer to treat it as a type of conduct for the good of the community. As the distribution of the work of content creators becomes increasingly granular in the CM process and as the industry adopts a more return-on-investment approach, the autonomy of the technical writer, which has always been championed as the mainstay that defines the humanities heritage of the technical writer, is seemingly at odds with the CM workflow and the expectations of the industry. The logical question then is: can we preserve this craft tradition in this age of CM?

The question central to this issue is what practical purposes technical writing should serve. What makes this a complex issue is that the academia and the industry define "practical" quite differently. In her seminal 1989 essay "What's Practical about Technical Writing," Carolyn Miller distinguishes between two senses of the "practical"—the *low sense* is concerned with getting things done efficiently and effectively while the *high sense* refers to "human conduct in those activities that maintain the life of the community" (p. 15). The problem with the low sense of the practical is that nonacademic rhetorical practices are inadequate while they serve as authoritative models. Technical writing, therefore, needs "a basis for evaluating a practice other than that of the practice itself" (Miller, p. 18). Rhetoric is practical knowledge (concerned with doing), rather than theoretical knowledge (concerned with knowing for its own sake) or productive knowledge (concerned with making). Rhetoric "emphasize[s] action over knowledge or production; rhetoric becomes a form of conduct" and is concerned not just with the useful but with the good (Miller, p. 22). As Miller has argued,

> understanding practical rhetoric as a matter of conduct rather than production, as a matter of arguing in a prudent way toward the good of the community rather than of constructing texts, should provide some new perspectives for teachers of technical writing.
>
> *(p. 23)*

Academics should know about nonacademic practices while also critiquing those practices.

Understood from this perspective, technical writing, even within the context of CM, still has a rightful place for the autonomy of the writer/content creator, however granular the distribution of work has become in the CM workflow and however complex the CM process has rendered technical writing due to the multitude of stakeholders involved. Technical writers in the CM context should maintain their autonomy for the good of the larger community rather than simply for the sake of the corporate community. As Bridgeford has argued in the Introduction, technical writing in the age of CM has become an act of true collaboration within a networked community, rendering their conduct a form of networked agency. Within this networked agency, all participants in the CM workflow—content strategists, information architects, writers, editors, content managers, etc.—contribute to this networked community and take responsibility for their individual and collective actions. Autonomy still counts, and so does the craft tradition.

Content-management-driven Pedagogy Conceptualized and Operationalized

Now the important questions: how does CM inform our pedagogy? What does a CM-driven pedagogy look like? What do we teach in our CM classes?

For technical communication instructors new to teaching CM, these can be formidable questions. We need to be realistic: if you are teaching CM for the first time, you should not expect students to walk out of your first course in CM ready to design or customize their own CMS. What we offer here is certainly not a teaching module that will produce CM experts. Rather, it serves to get your foot off the ground. We will discuss CM-driven pedagogy at both conceptual and operational levels.

Content-management-driven Pedagogy Conceptualized

At the conceptual level, a CM-driven pedagogy should focus on developing the right mindset in students, changing their perceptions, and guiding them to think of technical communication as a CM endeavor. On the conceptual side, we recommend that your CM course focus on the following:

1. *How digital information differs from print:* although the ubiquitous digital technology seems to warrant the assumption that students understand well how digital information differs from print, the not-so-infrequent cases when students turn in a digital project that clearly has employed the hardcopy design approach without an awareness of the differences in medium tell us that such discussions in your first CM class meetings are more than necessary. One possible activity that can help drive home the point about digital vs. print is to compare, for example, a hardcopy user manual with an online help tutorial. Discussions and assignments can encompass such issues as the content design and organization, medium of delivery, user expectations, context of use, etc.
2. *Types of CM:* students are not necessarily clear about the differences between CM, ECM, CCM, web CM, etc. A white paper on this topic could very well raise students' awareness of the important differences.
3. *The process of CM:* for students typically writing single assignments and projects, the complicated process of CM is often a foreign concept. A discussion of the CM process from planning to content creation, content storage, content output, and content delivery will give them a conceptual foundation about CM. An exercise that puts students into the role of the content manager for a small organization could force students to think about the big picture of CM instead of single communication products.
4. *Content strategy:* with a good understanding of the CM process, strategizing about CM will naturally come into play. Here it is appropriate to discuss the role of a content strategist on the CM team. An interview assignment of a corporate content strategist will give students a first-hand perception of what a content strategist does and how important this role is to the entire CM process.
5. *Stakeholders:* for the traditional single-document design approach, the concept of stakeholders is relatively simple, although deceptively so. Students typically think of the writer and the reader and the relationship between the two. With CM, the list of stakeholders instantly becomes numerous and complex and may involve such players as "technical communicators, information designers, UX specialists, content strategists, subject-matter experts (SMEs), technical translation/localization specialists, etc." (Batova, 2014, p. 326).
6. *Content credentialing:* this entails who can view, edit, distribute documents.

7. *Localization:* in many cases of CM, localization is an integral component. Although it is not within the scope of our discussion in this article, it is helpful for students to understand the variations of content based on where the audience is coming from, primarily job-related differentiation but also perhaps cultural.
8. *Workflow:* it is important for students to understand a document's planning, creation, design, and distribution processes. An assignment that asks students to design a workflow for a simulated case may very likely produce as many versions as there are students in the class, which will serve as a good illustration of the importance and significance of designing a sensible and logical workflow.

This is certainly not an exhaustive list, but a good understanding of these concepts should ultimately give students CM awareness and even the ability to produce a design schematic for a CM project.

Content-management-driven Pedagogy Operationalized

On the operational side, we can easily name hundreds of topics, but we think the following list will give students a good start in getting their feet wet in CM practice:

1. *Information architecture:* topics of teaching in this category may include content structure, menus, paths, etc.
2. *Screen template design:* for any technical communication student who wants to become a content manager, an understanding of how things work in the back end, including some coding knowledge, would be helpful. For starters, some practice in HTML and CSS would give them a much better sense of how the code works in the back end to produce the web pages.
3. *Document Type Definitions (DTDs):* an essential component of CM, DTDs inform the student of such concepts as content description, content classification, and content labeling. Digitizing a DTD will be a good exercise for students by having them create forms with, for example, PHP and MySQL.
4. *Prototyping:* both low-fidelity prototyping, such as simple rendition on word processors, and high-fidelity prototyping, such as wireframes and screen templates, will be especially meaningful practice.
5. *Usability testing and iteration:* usability is certainly not a CM-exclusive concept, so students should be fairly comfortable in conducting usability tests and then re-designing based on user feedback.
6. *Single sourcing:* an essential technology to CM, especially CCM, single sourcing will allow students to create drafts and archives not visible on the surface.
7. *Segmentation:* arguably a challenging task, the key to segmentation is a good understanding of the organization's content needs and content use. For example, segmentation for the purpose of translation and localization

will be very different from segmentation for content output targeting a domestic audience in that the level of granularity will differ.
8. *Inventory of existing CM software solutions:* a knowledge of what CMSs are available and their strengths and limitations will be very important for determining what software solutions will work best for an organization's CM endeavor.

Any CM-driven pedagogy must start somewhere. What topics to cover in your course will depend on your particular course goals and objectives.

An Example for Digital Content Creation

A great assignment to help novice technical writers come to understand the difference between paper and digital media is to have them produce the same content for a range of prevalent platforms: Instagram (image/infographic), YouTube (video), Twitter (140 characters), a blog post (images and text), email blast, a website (telescoping text, possibly long). Seeing how to play the same content out at different lengths in different media is an excellent way to better understand what technical writing is today.

Because digital content exists in a database, making informed decisions about document structure enables search and also allows for content to be repurposed and redeployed as new situations arise or as previous situations return. Thus, people new to technical writing need to learn about DTDs and they need to know how to define document structures.

Let us take as a concrete example a plausible structure for a DTD of an assignment. Let us say that the parts of all well-formed assignments are as follows:

1. Learning outcome: which of the class's stated learning outcomes does this assignment address?
2. Assignment objective: what will they learn? (And how might that learning be measured?)
3. Introduction (and motivation): what are they being asked to do, and why? Understanding an assignment's relevance to course objectives and, preferably, life beyond school motivates learning.
4. Instructions: what to do and, perhaps, how to do it.
5. Deliverable: what is to be turned in.
6. Examples: paradigms can be usefully instructive but also restrictive and or intimidating. If you do offer examples, offer a range of success and perhaps a cautionary tale.
7. Evaluation procedures and description of the deliverable the student is responsible for and how you will grade it.
8. Proportion of final grade.
9. Due date.

Given this structure, have students create a few assignments, say assignments that would help a novice writer practice style and writing for different audiences, to see how prescribing a structure both enables a complete workflow and also makes it possible to sort and organize content by means other than just what is in the document. Given this DTD, and a handful of instances of assignments so designed, you could sort assignments by learning outcome, due date, motivation, or instructions, indeed by any elements of the definition.

A related assignment would be to create a music playlist database that consists of metadata that would enable a person to discover music by means beyond author, title, genre, and so on. Ways like beats per minute or by theme—*show me all songs that reference Memphis*. A database is only as good as the metadata provided, and metadata schemas come from people who think about how the content can (and could) be described. This is a plausible job for the people who write the content, people who have a broad understanding of all information being offered by a company or corporation in its many different guises.

Thus, exercises in generating multiple ways of referring to the same object, sometimes called faceted classification, should be a part of any technical writing class. Given a photograph, for example, how many different ways can you think of to refer to it: point of view, geography (where), topography (the scene—beach, mountain, lake, etc.), time of year, time of day, objects in the foreground, background, color scheme, people or no people, individual or group. And finally, likely places in the content typically provided by the business where this photograph might be placed. This last "asset," the catalog of places for photographs, is a critical part of well-managed content workflow because photographs have huge click-through value. People are much more likely to read more than a headline if the headline is accompanied by an interesting image. One does not always have current "art," as the journalists used to call it, and technical writers are not often journalists, so a repository of well-identified "art" can be helpful. One can, of course, go to Shutterstock and the like, but then copyright is an issue and local content, pictures of people who actually work at your company, will always play better.

Audio files, podcasts, and interviews also need to be logged and tagged so that they can be used for their original purpose and snipped from for subsequent purposes. Perhaps something the CEO said at the annual meeting of shareholders could be worth using as a quote at the beginning of a new blog entry.

Conclusion

Under the heading of "old is new again," one of the most important roles for writers and public speakers is collecting, organizing, remembering, and reusing information from multiple sources. In the literate past, the commonplace book was the tool used for this purpose. It consisted of folios that could have pages added to them on which you might transcribe something that you witnessed (today perhaps a photograph), or something that you hear (today an audio file) or a quotation of something you read. Quotations were by far the most common entry. They were often organized by occasion and theme, making it easier for the collector to find what he needed for a sermon on the advent of Lent, for example, or what to say in a eulogy for a young child.

Today we call this practice "asset management," and knowing what is contained in the thousands of files on your webserver, where they came from, how old they are, what purpose and for what audience they were intended and how they might be retooled for other purposes is a critical role for content providers. Such a shift from "writing" to content/asset management dictates a reconceptualization of our traditional pedagogical practice. In this digital era where writing becomes CM and where writing is rendered a much more complex act within a networked community, our pedagogical practice needs to not only adjust accordingly but also do it in a critical manner.

References

99Robots. (2014). The difference between hosted and self-hosted website platforms. Retrieved August 6, 2019, from https://99robots.com/the-difference-between-hosted-and-self-hosted-website-platforms/.

Allen, R. (2012, July 26). Designing content workflow for your CMS. *Meetcontent.* Retrieved August 6, 2019, from http://meetcontent.com/blog/designing-content-workflow-for-your-cms/.

Association for Information and Image Management. (n.d.). What is enterprise content management (ECM)? Retrieved October 27, 2017, from www.aiim.org/What-is-ECM-Enterprise-Content-Management#.

Barker, D.(2016). *Web content management.* Sebastopol, CA: O'Reilly Media.

Batova, T. (2014). Component content management and quality of information products for global audiences: An integrative literature review. *IEEE Transactions on Professional Communication*, *57*(4), 325–339. doi:10.1109/TPC.2014.2373911.

Boiko, B. (2005). *Content management bible* (2nd ed.). Indianapolis, IN: Wiley.

Bollinger, I. (2014, September 5). When to choose a hosted versus open source CMS. Retrieved August 6, 2019, from https://trellis.co/blog/choose-hosted-versus-open-source-cms/.

Bussolati, M. (n.d.). 10 definitions of content strategy. Retrieved October 28, 2017, from https://bussolati.com/contentstrategy-definition/.

Churchville, F. (n.d.). Content management system. Retrieved October 27, 2017, from http://searchcontentmanagement.techtarget.com/definition/content-management-system-CMS.

Clark, D. (2008). Content management and the separation of presentation from content. *Technical Communication Quarterly*, *17*(1), 35–60. doi:10.1080/10572250701588624.

CMS Review (n.d.). Seven stages of the content lifecycle. Retrieved October 28, 2017, from www.cmsreview.com/Stages/.

Crouch, D. (2017, November 9). Hosted vs. non-hosted ecommerce solutions: Which is better? Retrieved August 6, 2019, from www.slatwallcommerce.com/resources/articles/hosted-vs-non-hosted-ecommerce-solutions-which-is-better/.

Gilbane. (n.d.). Component content management: How true CCM drives the most compelling content initiatives. Retrieved October 27, 2017, from https://gilbane.com/component-content-management-true-ccm-technology-drives-compelling-content-initiatives/.

Kanga Internet (n.d.). Content management systems: The history and the future. Retrieved October 26, 2017, from www.kangainternet.com.au/web-development/content-management-systems-the-history-and-the-future.html.

Kiwak, K. (n.d.). Content management. Retrieved October 27, 2017, from http://search contentmanagement.techtarget.com/definition/content-management.

Laserfiche. (n.d.). What is enterprise content management? Retrieved October 27, 2017, from www.laserfiche.com/ecmblog/what-is-enterprise-content-management-ecm/.

Levitt, J. (2011, December 18). When was the first web content management system (CMS) created? Retrieved August 6, 2019, from www.quora.com/When-was-the-first-web-content-management-system-CMS-released.

Miller, C.R. (1989). What's practical about technical writing? In B. E. Fearing & W. K. Sparrow (Eds.), *Technical writing: Theory and practice* (pp. 14–24). New York, NY: MLA.

OnBase. (n.d.). What is enterprise content management (ECM)? Retrieved October 27, 2017, from www.onbase.com/en/learn-ecm/what-is-ecm#.WfN9QRNSzq0.

Rand-Hendriksen, M. (2016). What is content strategy? Retrieved August 5, 2019, from www.linkedin.com/learning/ux-foundations-content-strategy/what-is-content-strategy.

Smith, H. (2014, June 12). What is content strategy? Distilled. Retrieved August 6, 2019, from www.distilled.net/resources/what-is-content-strategy/.

Tolvanen, P. (2013, November 11). Future, present and history of CMSs as told by CMS expert Deane Barker. Retrieved October 27, 2017, from https://northpatrol.com/2013/11/11/future-present-and-history-of-cmss-as-told-by-cms-expert-deane-barker/.

Trotter, P. (2008, February 4). Component content management: What is it and why does it matter? The Content Wrangler. Retrieved October 17, 2017, from http://thecontentwrangler.com/2008/02/04/component_content_management_what_is_it_and_why_does_it_matter/#.

W3Techs (2017). Historical trends in the usage of content management systems for websites. Retrieved August 6, 2019, from https://w3techs.com/technologies/history_overview/content_management/all.

We Make Websites. (2014). A hosted vs self-hosted comparison for e-commerce. Retrieved November 3, 2014, from, https://wemakewebsites.com/blog/a-hosted-vs-self-hosted-comparison-for-e-commerce.

Whitehead, J. (2014, January 2). The rise of Drupal and the fall of closed source. Retrieved August 6, 2019, from https://opensource.com/business/14/1/drupal-history.

Wikipedia. (n.d.). Content management. Retrieved October 27, 2017, from https://en.wikipedia.org/wiki/Content_management.

Winters, J. What is a content management system? Retrieved October 26, 2017, from http://bahai-library.com/what_is_cms.

2

CONTENT MANAGEMENT

Preparing Technical Communication Students for the Realities of the Workplace

Saul Carliner

CONCORDIA UNIVERSITY, MONTREAL, QUEBEC

Chapter Takeaways

- Link content management competencies to broader competencies to be developed in the curriculum.
- Each course focuses on specific content management competencies that relate to the broader competencies to be developed in the course.
- Use authentic assignments when developing content management competencies in courses: assignments that are representative of the work of practicing professionals.

Imagine you have just been assigned to teach a cross-listed undergraduate–graduate course in technical communication with the additional instruction, "Cover content management."

Soon afterwards, you encounter an online article by a newly graduated undergraduate in technical communication a few months into her first job. The author comments that

> technical communication is centered around the tools that you use to achieve more efficient writing and content management, like structured authoring and DITA [Darwin Information Typing Architecture]. But many companies often have a hard time building up technical writing teams that have experience using these tools.
>
> *(Heath, 2018)*

She then thanked her professor for teaching the tools (Heath, 2018).

As a result of reading that article, you assume that content management means teaching DITA and structured authoring. You even mention this when talking to Paul at a Society for Technical Communication (STC) meeting the following week. "Be careful," he warns, confessing that too much tools knowledge has stalled his career. Paul has about a decade of experience working for a medium-sized technical communication group as its "tools person." He ostensibly has responsibility for maintaining most of the programmer documentation for the company's oldest product, but he is best known within the organization for his mastery of production issues in the complex content management system. Paul can write routines and troubleshoot many problems, and people respect his tools knowledge. Because Paul is so knowledgeable with tools, management is reluctant to shift him into a writing assignment that would demand more of his attention, such as preparing documentation for a new product (which would be a full-time effort), much less a promotion to a role in content strategy—that is, planning the information would be provided to particular user groups of a product, the formats of that information, and looking at ways of linking material to avoid unnecessary duplication. As a result of his knowledge of tools for production, Paul feels his career plateaued years ago and seems stuck for the foreseeable future.

Paul's tale suggests that you should teach tools knowledge to a point, though you are not quite sure yet where to draw the line between enough and too much tools knowledge. So you contact Juliana, a technical communication consultant you know. She advises focusing on concepts of content management. Then she proceeds to tell you about an experience she had in the mid-2000s—the early days of content management. A large, regulated, multi-site company contacted her to help them with a technical problem. To comply with new regulations that the regulated documentation be available on several platforms—print, PDF, and mobile—the company tried to comply by developing custom software based on its Structured Generalized Markup Language (SGML) system, a late 1980s, early 1990s-era technology intended to be used with mainframe computers. The company had already sunk several hundred thousand dollars into the effort, but the software was still not working.

Although the company preferred that everyone use the same tools, it did not require it. And another site decided against investing in this effort. It chose, instead, to use an off-the-shelf product that cost significantly less than the not-yet-finished software development effort. Furthermore, after nine months of work, that site was already producing documentation that complied with the regulations.

When the prospective client explained the situation, Juliana said that was when she knew she did not want this assignment, saying that the managers had no idea what they were doing. They wanted her to save a failing effort because either they did not understand the technology or could not bring themselves to extricate themselves from a doomed project. Juliana added that the prospective client really needed a basic education in how the technologies worked so they could realize on their own that their project would not work, instead of relying on opinions of workers who were either equally uninformed or unwilling to concede defeat.

Three encounters. Three different directions for the course. What to do?

With three differing opinions, advice that might have seemed helpful individually just feels like the conundrum faced daily by technical and professional communicators: three subject matter experts providing three differing opinions about how to present content to readers. You have to make the decision on your own.

And just as the context of use guides the ultimate choice in presenting material to readers, so context also guides the choice in how to teach content management to students. This chapter sorts through these contextual issues so instructors can figure out how to approach content management in their courses. It starts at the end: the curricular context of addressing content management in technical communication education. That discussion considers the role of content management in developing students' competency as technical communicators. In the process of doing so, competencies in content management are suggested. This chapter continues by considering the competencies in content management that typical courses might explore and suggests how instructors might incorporate those competencies into their courses.

Competencies, Curricula, and Content Management

According to the general literatures on teaching and learning in higher education (Diamond, 1998) and instructional design (Merrill, 2002), effective instruction begins at the end: what should students be able to do when they complete a program of study? In the context of professionally focused programs, that involves preparing students for the workplace and to be competent within their chosen field. Competencies are central to this goal. They are "clusters of interrelated knowledge, skills, attitudes, and values necessary for performing [a job-related responsibility] effectively" (Institute for Performance and Learning, 2016, p. Introduction-1). The STC (2016) has identified a competency model for technical communicators. The STC Competency Model encompasses nine competencies:

- Project planning, which focuses on defining the publication process and scheduling a project.
- Project analysis, which focuses on defining the audience and context (but not tasks) of a project.
- Content development, which focuses on researching a topic, writing in genres, and observing copyright laws.
- Organizational design (not to be confused with the concept in human resource development, which has an entirely different focus), which focuses on structuring content and using genres to generate a document structure.

- Written communication, which focuses on writing in plain language, how to write different parts of a document, and how to write for specific emerging genres.
- Visual communication, which focuses on page and screen design, as well as using images to communicate ideas.
- Review and editing, which focuses on assessing content prepared by other technical communicators to ensure that it is clear, concise, and readable, and that the content conforms to established editorial, design, and technical guidelines.
- Content management, which focuses on the basic features of websites, collaborative writing, and content strategy.
- Production delivery, which focuses on awareness of the production process for content.

On the other hand, these nine competencies represent a small part of the complete set of competencies needed to work in technical communication. Part of that gap is the result of a competency model that only focuses on the unique competencies for technical communicators. According to the Building Blocks model proposed by the Competencies Model Clearinghouse (2018) and adapted in Table 2.1, all jobs have competencies in seven broad areas. Most of the competencies are common to all jobs.

The majority of the competencies in the Building Blocks model apply to all jobs, or all jobs in a particular industry. The most fundamental competencies shown in the lower three levels of the figure are ones required in all jobs. *Personal effectiveness competencies* refers to competencies such as interpersonal skills, integrity, and initiative. These competencies are developed from the youngest ages through a combination of home experiences, school, work, and life experiences. Students are expected to have developed these competencies before they start higher education, though certain co- and extra-curricular programs in higher education are designed to address deficiencies in these areas. The second set of competencies, *academic competencies* such as reading, writing, critical and analytical thinking, and basic computer skills, are developed through grade school and further enhanced through higher education. Students are expected to develop these competencies outside the scope of their major; either in high school, first-year programs, or academic preparation programs. The third set of foundational competencies, *workplace competencies*, such as team work, customer focus, and problem solving and decision making, are developed through a combination of schooling, life experiences, and work experiences. Students are generally expected to develop these competencies through extra-curricular activities and part-time jobs, but faculty often try to foster these competencies through group work (such as Thompson, 2001), service-learning assignments (such as Matthews & Zimmerman, 1999), and work placements like cooperative education and internships (such as Munger, 2006).

TABLE 2.1 Broad Competencies Areas (Competency Model Clearinghouse, 2018)

Management Competencies	*Occupation-specific Competencies*
Staffing	Such as the STC Competency Model
Informing	and
Delegating	additional competencies for content management
Networking	
Monitoring work	
Entrepreneurship	
Supporting others	
Motivating and inspiring	
Developing and mentoring	
Strategic planning and action	
Preparing and evaluating budgets	
Clarifying roles and objectives	
Managing conflict and team building	
Developing an organizational vision	
Monitoring and controlling resources	
Industry-sector Technical Competencies	
Identified by representatives of the industry, such as systems specialists and translation specialists	
Industry-wide Technical Competencies	
Identified by representatives of the industry, such as technical communication or training and development	

Workplace Competencies

Teamwork	Customer focus	Planning and organizing	Creative thinking	Problem solving and decision making	Working with tools and technology
Scheduling and coordinating	Checking, examining, and recording		Business fundamentals	Sustainable practices	Health and safety

Academic Competencies

Reading	Writing	Mathematics	Science and technology	Communication	Critical and analytical thinking	Basic computer skills

Personal Effectiveness Competencies

Interper-sonal skills	Integrity	Professionalism	Initiative	Dependability and reliability	Adaptability and flexibility	Lifelong learning

The two sets of competencies in the middle tiers of the Building Blocks model focus on working in particular industries like hospitality, information technology, and manufacturing. *Industry-wide technical competencies* refers to competencies needed within an industry segment, and usually developed through work experiences and job-related training. *Industry-sector technical competencies* refers to competencies needed within a particular part of an industry—such as the aircraft segment of the transportation industry—and developed similarly to industry-wide technical competencies. Employers and trade associations typically take responsibility for developing both of these sets of competencies.

The left side of the top tier of competencies in the Building Blocks model is *management competencies*, which refers to competencies for staffing, strategic planning, and monitoring and controlling resources, which are learned in formal education—higher education or training—and honed on the job. These types of competencies are significantly broader in scope and influence than the project management emphasized in many academic programs. Occupation-specific competencies, such as those needed to work as a surgeon, civil engineer, and technical communicator, only occupy the upper right block of the top tier of competencies in the Building Blocks framework.

Linking the Building Blocks model to the STC Competency Model, the majority—project analysis, content development, organization design, written communication, visual communication, and content management—are occupation-specific competencies. The remaining two—project planning and production delivery—partially address management competencies and, in the case of production delivery, some industry competencies.

The other part of the gap between the nine competencies in the STC model and the competencies needed to work in technical communication is that the STC competencies do not fully reflect the competencies needed in many jobs in the field. This situation is especially true of content management. The STC Competency Model limits content management competencies to describing features of a website and discussing techniques and practices of collaborative writing. By contrast, although the authors of this book did not collectively prepare our own list of content management competencies, based on a reading of those chapters, they might also include composing material for dynamic publication, preparing a content strategy that employs content management technologies, implementing a content management system, and interpreting reports generated by the systems for strengthening work processes and meeting users' needs. This expanded view of competence in content management extends beyond the use of websites (in fact, it barely includes them), operationalizes the concept of collaboration within the context of content management, and introduces tasks associated with content development tasks and the role content management plays in the management and evaluation of our work. Table 2.2 contrasts the content management competencies in the STC Competency Model and ones that might be needed in the workplace.

TABLE 2.2 Comparison of Competencies for Content Management

Competencies Defined by STC	*Competencies Needed in the Workplace*
• "Describe the basic features of a website and how to set up and use websites in the workplace. • "Discuss techniques and practices of collaborative writing and content strategy" (STC, 2016, p. 4).	• Describe the capabilities of content management software for (1) managing the content development process; (2) dynamically publishing content in a variety of media; (3) facilitating translation of content; (4) providing metrics on the use of content; and (5) archiving content. • Compose material for dynamic publication in a variety of media by (1) writing structured content; (2) tagging content so that it can be selected for publication; (3) addressing potential differences in audiences and media when writing a single topic; (4) ensuring the consistency of technical material across topics; and (5) ensuring editorial and design consistency across topics. • (Advanced) Prepare a strategy for communicating a body of content employing content management technologies while addressing the needs of the targeted audiences as well as the technology, editorial, and project constraints imposed by the client requesting the content. • (Advanced) Given a set of requirements regarding publishing needs, internal work processes, and anticipated developments in an organization, implement a content management system that addresses those needs in that organization. • (Advanced) Use the reports generated by a content management system to strengthen internal work processes and expand the reach of the content by better meeting the needs of users.

One reason the STC model does not encompass this definition of content management competencies is that the STC model was designed to be testable for credentialing. STC bases its certification off of the model and that mode, therefore, must reflect current work assignments of the majority of technical communicators. Prior studies have shown that many technical communicators—somewhere between 40 per cent and two-thirds—do not work with content management or only do so in limited ways (Carliner & Chen, 2018a; O'Keefe, 2009). That does not mean that content management is unimportant to teach; it just means its role in the everyday life of working technical communicators varies widely.

Furthermore, the presence of a single competency model for technical communication suggests that the job has a singular identity.

But that suggestion is wrong.

Actual technical communication jobs vary widely. That is partly because the organizations hiring technical communicators have differing needs. Some differing needs result from the industry in which the employer operates and the resulting communication needs (for example, technical communicators in engineering firms generally work on reports and proposals, while technical communicators in the defense industry generally work on a variety of user and maintenance documentation), as well as whether technical communicators primarily serve internal or external audiences (for example, technical communicators working in IT departments generally serve internal audiences while those in product development groups generally serve external audiences). Further differentiating the jobs of technical communicators are the sizes of the groups in which they work, which range from lone writers (that is, a single technical communicator who performs all tasks) to large operations exceeding 50 technical communicators and characterized by specialization of tasks and two or more layers of management. So, the only assumption one can make when hearing that someone works as a technical communicator is that they play a role in transferring knowledge from those who know to those who need to know.

In fact, specialization is central to understanding the work of technical communicators in general, and the impact of content management on that work in particular. Rather than referring to a single job role, research such as the 2018 Census of Technical Communicators (Carliner & Chen, 2018a) suggests that technical communication work generally falls into four broad job roles regardless of other contextual issues in the work environment: a planner/designer role, a creator role, a production role, and a management role.

The planner/designer role involves planning and design, in which the technical communicator has responsibility for "solving the complex communication problems presented by project sponsors (that is, the programmers, engineers, marketing professionals and others who hire us to communicate their technical content with a group of designated users)" (Carliner, 2001, p. x) and developing plans for the solution, which usually entail a number of related materials. This role always involves an investigation phase, called audience and task analysis, and also called a needs analysis, to identify what needs to be communicated, to whom, and any constraints on the communication effort (such as a requirement to use a particular technology to create the materials or limitations in the reading levels of the audience). Designing involves conceptualizing how the material will be communicated to its intended audiences and results in plans, which could be as simple as outlines for a library of documentation for a hardware product or as complex as an integrated campaign of technical documentation, instructional materials, and marketing materials crossing several media, involving the preparation of a content inventory, taxonomy, and integration into complex content management systems. The scope of the work is moderate to broad, affecting departments, part or all of entire product lines, and perhaps divisions or corporations. Planners/designers lead the development of design and editorial guidelines and the selection and application of technologies for the others working in their work teams or projects. This job role has had several titles over the years, including planner, information designer, and, more recently, content strategist (Carliner, 2001; Clark, 2016).

The creator role involves preparing one or more of the communication pieces outlined in the plans. Creating involves in-depth research of the technical material, considering how the intended audience might use it, and writing the material outlined in the plans. When technical communication first formally emerged as a field in the 1940s and 1950s, that task typically involved writing a manual of some sort, such as user guide, reference manual, or parts catalog. As online materials emerged, some of that material took new form online as online help, frequently asked questions (FAQs), and technical websites and knowledge bases. With the rise of content management systems, in which pieces of information can be combined on the fly, creating often involves writing topics on a particular subject, though the topics can take the forms of descriptions, procedures, and other microgenres familiar to those who wrote manuals. The scope of the work is limited to moderate, affecting a part of a product line, area of service, or organizational unit; some creators participate in the establishment of guidelines and the selection of technology affecting their work; others follow guidelines and use technology selected by others. Once universally called a technical writer, this role has been given a variety of names in the past few decades, including technical communicator, information developer, product information specialist and, more recently, content developer (Carliner, 2001).

The third role is a production role, which is primarily responsible for readying draft material for publication. In the early days of technical communication, when organizations primarily published material in print, this involved taking typed or handwritten manuscripts from the creators (writers), typesetting the material, laying it out, and inserting images, and preparing a package suitable for transmission to a printer. The rise of word processing and desktop publishing in the 1980s and early 1990s significantly reduced the need for print production specialists, because creators could increasingly type their own manuscripts and prepare them for publication. When online help, web publishing, and content management emerged in the 1990s and 2000s, the systems were designed so that creators could perform as much of their own production as possible, although some assistance might be needed with more complex production tasks. When organizations required specialized production assistance to produce printed materials, the role was called an editorial assistant or production assistant. As creators could increasingly perform their own production tasks, the role has become increasingly less common. But because some production assistance continues to be needed, the responsibilities are increasingly being included in the responsibilities of creators. In addition to the roles of production assistants in preparing materials for publication, technical editors have traditionally played a role in readying materials for publication. In the days of print publication, technical editors would mark up manuscripts of materials to indicate the levels of headings, provide instructions for formatting tables and figures, ensure that figure and table captions were properly inserted and, in some organizations, oversaw the preparation of indexes. Authoring tools let creators handle these tasks on their own, reducing demand for the production responsibilities of technical editors. As a result of incorporating production assistant and technical editor responsibilities into the everyday responsibilities of creators (writers), some creator jobs might look more like production jobs because they involve bits of both roles (Carliner & Chen, 2018a). The scope of the work is limited. Most production assistants are guided by design and editorial guidelines established by others and use technologies selected by others.

The last role is that of a manager, who has responsibility for overseeing the work of a group as well as personnel responsibility for the people who work there, including hiring, firing, establishing performance objectives, conducting performance evaluations, providing ongoing coaching, and supporting the development of workers. Although the nature of the work overseen and skills needed by workers have changed over the past few decades, and the management of projects has grown in importance, the job title has remained the same. The scope of the work is moderate to broad, affecting departments, part or all of entire product lines, and perhaps divisions or corporations. Managers oversee the development of design and editorial guidelines and the selection and application of technologies for the teams they supervise. Some managers take a hands-off approach, letting others in the team conduct the research and make recommendations. Other managers play a more hands-on role in these decisions.

Each role involves a different type of engagement with content management. The planner/designer, for example, leads decision making regarding the use of content management and its application to particular projects. The creator applies those decisions and develops content that conforms to those guidelines. The producer ensures that completed content conforms to any standards and appears as intended within the content management infrastructure. And the manager approves all decisions regarding content management, requests resources to implement the decisions and ensures that the staff implements them. Table 2.3 describes the specific ways that individual roles engage with content management.

The general competencies needed by technical communicators and the specific roles addressed by the program guide the creation of curricula for professional and technical communication programs. For example, some programs focus on particular roles, such as technical communication managers and content strategists, while other programs take a more generalist approach. Specialist programs might develop competencies with content management suited to the

TABLE 2.3 How Different Roles Engage with Content Management

Role	*Engagement with Content Management*	*Specific Tasks with Content Management*
Planner	Leads decision making regarding the use of content management and its application to particular projects	• Participate in selection of the system. • Determine how to exploit the affordance of the technology given the context in which it will be applied. • Develop design templates/formats/models that others will follow. • Establish or lead development of guidelines for others to follow.
Creator	Applies decisions about content management and develops content that conforms to those guidelines	• Develop content within the system and that conforms to the established guidelines, including guidelines for re-using material. • Adapt writing practices to the templates/formats/models. • Participate in the development of guidelines. • Manage different versions of the same document. • Ensure smooth passage or work through the review, revision, and production cycles.
Production assistant	Ensures that completed content conforms to any standards and appears as intended within the content management infrastructure	• Produce content within the system and that conforms to the established guidelines. • Manage different versions of the same document. • Ensure smooth passage or work through the review, revision, and production cycles.
Manager	Approves all decisions regarding content management and ensures that the staff implements them	• Represent the technical communication group in the selection of the system. • Establish the processes guiding the work of technical communication groups. • Ensure efficient use of resources, including the re-use of content. • Ensure that guidelines are established and followed.

role of interest, while generalist programs might take a broader approach. Similarly, groups that offer programs at varying levels—undergraduate, graduate, certificate or diploma, and continuing education—might consider the nature of competencies appropriate to the level of the program. For example, according to some models (Carliner, 2001; Willis, 1990), undergraduate education is intended to prepare students to do and graduate education is intended to prepare students to lead or manage. As a result, an undergraduate program might emphasize competencies related to using content management while a master's program might emphasize competencies related to leading or managing a content management effort. Graduates of all of these programs would leave with broad competency in content management, but the specific nature of that competency would vary depending on the roles for which the program prepares students upon graduation. Ultimately, decisions about the focus of programs and the resulting competencies to be taught are curriculum decisions.

Integrating Content Management into Particular Courses

Curricula only define the general nature of competencies to be developed; students develop these competencies in individual courses. Whether undergraduate or graduate level, degree programs in professional and technical communication programs typically have eight to ten required courses. Some focus on communication skills, some on production and technology skills, and some on management skills. Content management is relevant to all of these, but because of its complexity, the subject is differently relevant to all these courses. A communication course might focus on structured writing practices, a production course might focus on standards and coding, and a management course might focus on the impact of content management on work flows. Different types of courses, then, develop different types of content management competencies.

When assigned to teach a course, then, instructors might first consider the nature of the course subject, how it prepares students for the intended role upon graduation, and the competencies that the curriculum planners hoped it would develop. With this overview, instructors might then consider how content management connects with the other competencies developed in the course. Then, instructors might consider how to integrate content management into the course. Table 2.4 identifies some common courses, suggests their likely role in a curriculum and competencies in content management that instructors might address in those courses, and proposes assignments that might develop those competencies. Such assignments are also intended to make the material relevant to the future work of students.

Note that no course in Table 2.4 covers all competencies in content management. Instead, each covers content management competencies pertinent to the other competencies that comprise the main focus of the course, and rely

TABLE 2.4 Possible Ways to Integrate Content Management Competencies into Common Courses in Professional and Technical Communication Programs

Course	*Likely Role in the Curriculum*	*Competencies in Content Management to Address*	*Possible Assignments that Might Develop Those Competencies*
Professional and Technical Writing 1 (an introductory technical writing course for majors, which often focuses on definitions, procedures, and other explanatory material)	Introduce students to the genres of technical writing	Develop content within the system that conforms to the established guidelines	• Submit work through a Learning Management System (simulates submission in a content management system). • Follow file-naming conventions when submitting work. • Use named styles in *Microsoft Word*.
Professional and Technical Writing II (a course that often focuses on projects like user guides, online help, topic-based writing, and tutorials)	Prepare students for basic professional assignments in the field	• Develop content within the system and that conforms to the established guidelines. • Adapt writing practices to the templates/formats/models. • Participate in the development of guidelines.	In addition to continuing the requirements of the introductory course, also consider: • Preparing material that conforms to a particular template or style (structured writing). • Devising consistent terminology. • Assigning students to write related topics but in such a way that individual styles are not detectable. • Revising material to minimize impact on translation. • Describing content management, its components, and their impact on the

(Continued)

TABLE 2.4 (Cont.)

Course	*Likely Role in the Curriculum*	*Competencies in Content Management to Address*	*Possible Assignments that Might Develop Those Competencies*
Production course (a course that focuses on website production and desktop publishing skills)	Prepare students to produce materials, especially web-based and video materials	• Produce content within the system and that conforms to the established guidelines. • Develop templates/formats/models.	production of communication products. • Producing material within existing templates. • Preparing material that works with a given standard and without errors (such as DITA or the e-learning standards xAPI and SCORM). • Developing templates for others to follow.
Project Management	Prepare students for the business context of work projects, including establishment of—and conformance to—budget, schedule, and quality commitments	• Represent the technical communication group in the selection of the system. • Ensure that guidelines are established. • Ensure that guidelines are followed.	• Describing content management, its components, and their impact on the workflow for designing and developing communication products. • Choosing a simple content management system for a small organization. • Preparing for an upgrade to a content management system midway through a project and that could affect its workflow. • Developing technical guidelines for a project to ensure that all material can be used within a given content management system.

on assignments that mimic parts or all of real-world assignments for professional and technical communicators. These are called authentic assignments because they are representative of the type of work students are likely to encounter in the real world. For example, one of the suggested assignments in Professional and Technical Writing I is using named styles in *Microsoft Word*. Named styles are central to professional practice as they let users import and export material from one system to another with little to no loss in formatting, and many employers assess workers' adherence to this practice. Using named styles can be a stand-alone assignment. But more realistically, instructors might incorporate it into a broader assignment like writing a procedure. Procedures use many named styles, such as different levels of headings and different types of lists, and this provides a good context for teaching and assessing the competency. When marking the procedure, instructors can also assess the extent to which students mastered named styles by requiring that students submit *Word* files and checking formatting when marking the rest of the procedure. Similarly, students in a production course would still produce digital materials, but might develop and use templates for component content management systems rather than web content management systems.

The Advantages of Focusing on Competencies

This approach of focusing on competencies when teaching content management offers many advantages to instructors. The first is that it frees them from a false assumption that they must comprehensively teach content management in a single course. Content management is broad enough that this is probably not feasible anyway. The second is that, if the instructor has a limited background in content management, it helps focus their own preparation to teach the subject. An instructor would find out how content management affects the material in the course and just master those particular aspects of content management. Third, this approach allows students to develop competence in content management in each course for which the technology is relevant. Furthermore, if the same competency is relevant to two courses, the competency can be addressed in both courses. In some cases, the competency might be addressed in an identical way in both courses, and in others differently. That depends on the nature of the content and is ideally coordinated among instructors so that students sense the differences among the courses instead of perceiving an avoidable redundancy in teaching. Fourth, this approach also validates the role of content management in the work of professional and technical communicators: as a tool that supports the work. By maintaining the focus on content management as a tool, such an approach also affirms the primary roles of communication and awareness of the technology being communicated (which is almost *never* content management) in the work of professional and technical communicators.

References

Carliner, S. (2001). Emerging skills in technical communication: The information designer's place in a new career path for technical communicators. *Technical Communication, 48*(2), 156–175.

Carliner, S., & Chen, Y. (2018a). What technical communicators do. *Intercom, 65*(10), 10–16.

Clark, D. (2016). Content strategy: An integrative literature review. *IEEE Transactions on Professional Communication, 59*(1), 7–23.

Competency Model Clearinghouse. (2019) Building blocks model. Retrieved March 15, 2019, from www.careeronestop.org/competencymodel/competency-models/building-blocks-model.aspx

Diamond, R.M. (1998). *Designing and assessing courses and curricula: A practical guide. Jossey-Bass Higher.* San Francisco, CA: Jossey-Bass.

Heath, K. (2018) Training new hires in technical communication. *Scriptorium.com* Retrieved December 28, 2018, from www.scriptorium.com/2018/07/17514/

Institute for Performance and Learning. (2016). *Competencies for performance and learning professionals.* Toronto, ON: Institute for Performance and Learning.

Matthews, C., & Zimmerman, B.B. (1999). Integrating service learning and technical communication: Benefits and challenges. *Technical Communication Quarterly, 8*(4), 383–404.

Merrill, M.D. (2002). First principles of instruction. *Educational Technology Research and Development, 50*(3), 43–59.

Munger, R. (2006). Participating in a technical communication internship. *Technical Communication, 53*(3), 326–338.

O'Keefe,S. (2009.) *The State of Structured Authoring in Technical Communication.* Research Triangle Park, NC. Scriptorium.

Society for Technical Communication. (2016) *Certified Professional Technical Communicator (CPTD) study guide.* Arlington, VA: Society for Technical Communication. Retrieved March 15, 2019, from https://mk0julystc8rvl6sm3ml.kinstacdn.com/wp-content/uploads/2017/09/cptcstudyguide.pdf

Thompson, I. (2001). Collaboration in technical communication: A qualitative content analysis of journal articles, 1990–1999. *IEEE Transactions on Professional Communication, 44*(3), 161–173.

Willis, Verna. 1990. Presentation to the Atlanta Chapter of the International Society for Performance Improvement. March, 1990.

PART II
Teaching

3

TEACHING CONTENT STRATEGY IN TECHNICAL COMMUNICATION

Liza Potts

MICHIGAN STATE UNIVERSITY

Laura Gonzales

UNIVERSITY OF FLORIDA

Chapter Takeaways

- Training students to work as content strategists requires interdisciplinary collaborations among practices traditionally located in technical communication, professional writing, and experience architecture.
- Content management is a component of a well-designed content strategy, and both content management and content strategy are deeply rhetorical practices that are natural extensions of students' expertise in professional writing and experience architecture.
- A landscape analysis, tone and style guide, and content strategy document are three practical assignments that can help students practice contemporary content strategy techniques.
- One of the most important jobs of a technical communication instructor is to stay up to date with, if not one step ahead of, best practices, thought leadership (i.e., informed opinions and authorities in their specific fields), and technologies in industry. For this reason, instructors should update syllabi, assignments, and coursework on an annual, if not biannual, basis to continue sharing and learning current practices with students.

This chapter outlines approaches for teaching content strategy, a repeatable system defining the entire editorial content development process for a website (Sheffield, 2009). This work is done by content strategists, who develop systems for organizing, recontextualizing, disseminating, and managing digital content (Wachter-Boettcher, 2012). In describing these teaching approaches, we build upon previous work in content strategy that describes the shifting activities and training requirements for content strategists in industry (Hackos,

2002; Sheffield, 2009; Kissane, 2011; Wachter-Boettcher, 2012; Andersen, 2014; Batova & Andersen, 2016). By outlining teaching approaches, we work to describe and expand upon the relationships between content strategy and technical communication in both professional spaces and academic programs (e.g., Halvorson & Rach, 2012; Gonzales, Potts, Hart-Davidson, & McLeod, 2016).

In this chapter, we describe three assignments to help guide newer technical communication instructors in understanding how to apply these ideas in the classroom. The first assignment focuses on a landscape analysis, a research method for understanding how competitors and collaborators deliver content by way of websites, social media campaigns, etc. The second assignment is a tone and style guide, which allows students to take their research findings and use them to work through issues of voice and visuals for content campaigns. The third and largest assignment is a content strategy document. Through these interconnected assignments, new instructors will gain a clearer understanding of ways in which we can apply the theories and methods we learn in the classroom to these kinds of organizational deliverables.

Our goal with this chapter is to support instructors who teach content strategy in the technical communication classroom. As content strategy grows as a profession, and as technical communication continues shifting to include contemporary approached in content strategy, it is important for our field to continue to update, expand, and include leading, innovative research and best practices in the classroom. Throughout the rest of the chapter, you will find outlines, resources, and assignments to help prepare you to take on this work.

Approaches for Teaching Content Strategy

Content strategy involves more than just managing digital information, encompassing both "an organizational vision for information and an action plan for achieving it" (Batova & Andersen, 2016, p. 2). Content strategy "also involves the branding (such as creation and sustainability of a digital identity), organization (such as tagging, storing, and accessing information), management (such as the oversight and inventory of content), and delivery of content within and outside organizations" (Gonzales et al., 2016, p. 2). Thus, content strategy requires organizations to consider how content is being delivered across audiences and platforms, and how the image of the organization is being represented and adapted across media. Training students to work as content strategists requires interdisciplinary collaborations among practices traditionally located in technical communication, professional writing, and experience architecture (Halvorson & Rach, 2012; Gonzales et al., 2016; Potts & Salvo, 2017).

The assignments we describe in this chapter are taught in a course that serves both Professional Writing and Experience Architecture undergraduate majors. According to Potts and Salvo (2017), experience architecture is the strategic deployment of mediated systems, resulting in a designed capability for those using the systems to communicate. The Experience Architecture major at Michigan State University (where these assignments are implemented) is designed for Humanities students interested in gaining advanced skills in designing and developing experiences for people in digital and physical environments. Acting as a bridge between professional writing and experience architecture, the purpose of the assignments described in this chapter (and of the content strategy course we describe more broadly) is to help Experience Architecture students think about content and to help Professional Writing students develop strategies for working with developers and user-experience researchers. In this way, the assignments and course we describe in this chapter echo the expanding connections between technical communication, professional writing, and experience architecture now common practice in industry.

The specific learning goals of the assignments we describe in this chapter find a balance between theory and practice, between concepts and *techne* (i.e., practical strategies for building). Through these assignments, students have an opportunity to:

- Develop core competencies of content management for single sourcing, separating content from presentation, and distributing editorial roles among teams.
- Learn and practice the genres and tools crucial to content management.
- Learn and practice research methods for assessing the communication needs of organizations to recommend effective and sustainable content strategies.
- Situate content strategy as a component of user-centered design.
- Frame content strategy as a rhetorical practice and a professional writing craft.

The primary objective is to frame content management as a component of a well-designed content strategy, and to see both as deeply rhetorical practices that are natural extensions of their expertise in professional writing and experience architecture.

In short, the activities we describe in this chapter reflect a shift in focus to the *strategy* component of traditionally understood content management work, emphasizing the situational awareness and the cultural research required to effectively assess an organization's needs, understand its practices, and propose strategies that fit those cultures so as to be appropriate and sustainable. Content *management* is thus situated as one component of an effective *strategy*. In developing content strategy assignments, we seek to make visible how contemporary theories of technical communication are practiced in industry and to continue bridging relationships between content strategy and technical communication.

Relationships between Content Strategy and Technical Communication

As Tracy Bridgeford describes in the introduction to this collection, changes in content management (including content management systems (CMSs) and single-source technologies) have changed and are changing our writing and teaching practices in technical communication. As Andersen (2014) elaborates, "Technical communication (TC) practice has undergone what Hackos (2002) and Dicks (2009) have called a 'seismic shift'"; this shift moves away from a document-based to a topic-based approach to technical communication, rendering new processes, methodologies, and technologies "that enable content to be manipulated at a granular level" for more expedited development, reuse, and dissemination (Andersen, 2014, p. 116). Through this shift, approaches to technical communication practice and pedagogy should increasingly encompass successful content strategy.

Yet, as Andersen (2014) further explains, when the field of technical communication continues

> articulating and theorizing trends, methods, and technologies, we tend to situate our discussions within the existing scholarship rather than the larger CM discourse actively shaping CM practice. Our field's long struggle to keep pace with industry trends has made articulating a bigger picture view of CM diffusion difficult, but we also have not done a good job directly engaging in the robust and extensive CM conversations taking place outside of the academy.
>
> *(p. 116)*

The same is true, as Bridgeford clarifies in her Introduction, for pedagogies in technical communication, which need to catch up to emerging practices in content strategy.

Current literature on content strategy portrays a different role for the professional communicator than earlier literature in professional and technical communication. For example, Wachter-Boettcher's (2012) *Content Everywhere: Strategy and Structure for Future-ready Content* is a pragmatic look at specific locations and services that content managers are often expected to utilize, particularly diverse types of structured information, content models, and markup. One key difference in current content strategy literature pertains to the difference in the nature of the work: between being in charge of a publication and being in charge of a website. Being in charge of a publication involves the creation of content for a single site. By contrast, being in charge of a website involves organizing, recontextualizing, disseminating, and managing content.

Rather than serving as the author of an individual work or section of content or the manager of a project designed by someone else, the literature on content strategy portrays the professional communicator as a leader who establishes the strategy. For example, in *The Web Content Strategist's Bible: The Complete Guide to A New And Lucrative Career For Writers Of All Kinds*, Sheffield (2009) calls the role an editor-in-chief. Sheffield sees the role as the person responsible for overseeing all content requirements and creating all content strategy deliverables (such as content audits, gap analyses, and metadata frameworks). Other authors concur. Halvorson and Rach (2012) describe content strategists as the people responsible for overseeing the success of all content initiatives. Kissane (2011) explains that content strategists have to understand and facilitate both user goals and user behaviors, hence overseeing all aspects of content creation and dissemination. Sheffield (2009) chooses the term "editor-in-chief" because he is emphasizing the strategists' role in all aspects of the content, not just the initial creation.

Drawing on these connections between academic and industry practice in technical communication and content strategy, the assignments we share in this chapter present the role of the content strategist as editor-in-chief. This role means that students were not asked to necessarily develop new content for their clients but were instead asked to use research with their client to develop a strategy for creating, organizing, and disseminating current and future content.

Assignments

The detailed examples of three assignments that follow include a final project meant to guide the students towards the major goal of building a content strategy. The assignments we present in this chapter aim to continue bridging the gap between academic programs in technical communication and emerging industry practices, including advanced readings from academic researchers and deep dives in industry for honors/grad students in the class. Primary readings used to frame the assignments include *Content Strategy for the Web*, by Kristina Halvorson and Melissa Rach (2012), *Content Strategy Toolkit* by Meghan Casey (2015), *Going Responsive* by Karen McGrane (2015), and *Managing Chaos: Digital Governance by Design* by Lisa Welchman (2015). These books were selected because they represent powerful but approachable forms of contextual inquiry designed to develop a situational awareness that technical communicators, content strategists, professional writers, and user-experience architects instinctively know they need to successfully develop a product.

Landscape/Competitive Analysis

The Landscape/Competitive Analysis project is built to ease students into content strategy work by helping them learn about the form and content of different digital genres. It is a short project that appears early in the semester and only runs for two weeks. This project allows students who lean towards usability and information design to strengthen their skills and understand the importance of content in user experience, and it provides novices with the opportunity to learn about the connections between usability and content strategy. The assignment language describes this assignment in the syllabus:

> Project 1: Landscape/Competitive Analysis (Individual or Group Project)
> For your first project, you will conduct a landscape/competitive analysis. You will learn how to assess, compare, and contrast the ways in which content is organized and presented to various audiences. This analysis will help you begin to understand content elements, types, and objects. In this description, I specifically mention websites, but your team might want to analyze software, apps, digital magazines, or other kinds of digital content.
>
> **Goal:** Learn how to conduct a landscape/competitive analysis and report findings.
>
> **Deliverables:** Final report. Sending reflection individually.

Students are instructed to assess the ways in which content is written, designed, and provided for users and participants in digital spaces. The way we describe it to students, a landscape analysis is typically done for either non-profits or organizations not in competition with one another (e.g., the Philadelphia Orchestra website and the Los Angeles Philharmonic website). A competitive analysis typically looks across a channel or genre to assess competition in a given market (e.g., Netflix's TV system vs. Hulu's TV system, GM's Infotainment System vs. Ford's Infotainment System). Put simply, this analysis allows students to compare and contrast what works and what might need to be improved upon competing (competitive analysis) or similar (landscape analysis) digital platforms, apps, websites, etc.

As the Landscape Analysis assignment for this course is given only a few weeks into the semester, students are often nervous about the Landscape Analysis project and its goals. Walking the class through several examples of digital spaces—websites, screenshots of tablet interfaces and phones, etc.—and discussing different kinds of heuristics is critical to setting up this project. These heuristics can start with a discussion based on the US Government's usability website (usability.gov, 2018) and their reference to the set built initially by Jakob Nielsen (1994). We ask students to consider the tone and style used across the genre, encouraging them to think in terms of describing issues using terminology from Robin Williams' (2015) CRAP principles (Contrast,

Repetition, Alignment, Proximity). We then typically have students come up with their own set of heuristics that might be more specific and in tune to current issues. For example, much of our discussions in 2017 and 2018 focused on dark patterns (Greenberg et al., 2014; Brignull, 2018; Trice & Potts, 2018) and why we would not want to create them. As Trice & Potts (2018) explain, dark patterns are

> a user experience crafted to trick the user into performing actions not in the user's own interest (Brignull). Experts have looked at dark patterns in relationship to user interface (Brignull), physical proximity (Greenberg et al., p. 2), and social capital (Lewis 119).
>
> *(para. 13)*

We then use this collection of heuristics to walk through some examples to help students become more comfortable with doing the larger assignment, asking them to think of guiding questions such as "What works? What needs improvement?" and slowly getting them to start thinking about questions (such as "How or why does it work?") that will help them with their next projects.

Providing students with a template to get started and sharing past student (and even your own) examples is critical to modeling what these deliverables look like in practice. We emphasize the need to follow principles of CRAP as they design their slide deck, explaining how important it is to create a professional-looking deliverable as a persuasive tactic. This move allows students to see the connection between the genres they are analyzing and the genre they are working in. As instructors, we have found that using slides (*Google Slides*, *PowerPoint*, *KeyNote*, *Impress*, etc.) as a deliverable mode helps students tremendously. First, it helps them understand that slides are a mode of communication for practitioners. Second, it constrains them to a certain amount of material for the assignment (they can only fit so much on a slide and their examples model that for them) so as to help them learn how to time-box their projects (a critical skill for the final project). Finally, it challenges them in ways that they are not used to with regard to presentations. By telling students that the presentation must "stand alone," we are signaling to them that sometimes (heck, often) their deliverable is written for one audience but then viewed and used by a different audience.

Tone and Style Guide

The Tone and Style Guide assignment asks students to develop a document that will help content writers create useful content for an organization. This project is given more time for completion, as it requires more research and a more robust deliverable of roughly ten pages. As part of this project, students engage with a local client to understand their content process, governance, materials, and output. This project allows students who lean towards editing and publishing to expand their learnings and act as experts on the team, changing the dynamic from the first assignment that focused more so on usability and technical expertise. The assignment language is used to describe this assignment in the syllabus:

> Project 2: Content Style Guide with Content Models/Templates (Group Project)
> On websites, mobile apps, and desktop software, content is written, designed, and managed through the use of various kinds of tone, style, and templates. To create these guides, content strategists work with stakeholders to decide on the size of headings, the font choice for the text, the amount of content in each section, and the placement of images. They build templates and content models and create style guides. Part layout and design, part information architecture, and part branding, these guides help businesses meet their goals through content.
>
> **Goal:** Conduct research to understand an organization's content genre, and build a basic style guide, content model, and set of templates for a variety of platforms.
>
> **Deliverables**: Style Guide, Content Model, and Templates.

At the beginning of this project, we share multiple examples of content strategy guides, one-pagers, websites, and documents to help students understand the scope and diversity of these deliverables in practice. We explain that this project is part of the larger content strategy project, and as the semester progresses, we emphasize how the Tone and Style Guide is roughly one-third of that deliverable. The Guide should be a natural milestone for the team as they look towards building a more robust final deliverable. Sharing examples of the final Content Strategy document with them allows students to see how this Guide fits in with that larger deliverable.

Along with the Content Strategy document, we also share examples from industry. Some of these examples are one-pagers and some of them are much longer. Some of them include process information, and some of them do not. Across all these examples, we look for ones aimed at employees in organizations where they need to produce content. That is to say, we emphasize how not all of the "writers" in an organization are actually writers by trade; across an organization, everyone can and often will write material for the public. The job of the content strategist is to create guidelines for these writers to follow.

As students write this document, we point out the need to create a set of positive and negative examples for their organization to follow. This move helps students start to see how they are setting policy and creating governance for the organization through this Guide; we want them to be able to move beyond writing and into strategy. For example, if they want to set policy about how to write on Twitter, Instagram, Facebook, etc., they need to provide working models for their organization. We explain how these examples can help them show how to apply their instructions about tone and style, noting the textual and visual features of digital spaces.

Content Strategy Document

The Content Strategy Document assignment is the largest assignment of the semester. The Content Strategy is a robust, 30–35 page (with appendices) living document meant to contain information for running the content strategy of an organization. It contains information on tone, style, governance (including any process diagrams), privacy/security, templates, search engine optimization (SEO) suggestions, content models, crisis communication, and examples. It is meant to be used by the content strategist and deployed as needed. By the end of this assignment, both kinds of students (those who lean towards usability and design and those who lean towards editing and publishing), will gain insights into how the other works and why their work is valuable. This final project gives students the opportunity to pull together their course and program learnings; while the course was not created as a capstone, it has had that effect for several students. Indeed, many students have commented on how this document becomes an integral part of their final portfolio as they head out to the job market. Below is the language used to describe this assignment in the syllabus:

> Project 3: Content Strategy (Group Project)
> This project will help you learn how organizations create, organize, and sustain content. Or how they don't—and the consequences of it. Whether you are investigating a project in the library, the local animal shelter, or a high-tech start-up, you will consider several research questions: What kind of website, app, or mobile site are they maintaining? Who is their audience? Who are their writers? How are they managing the content workflow? Is their content sustainable? How are they handling governance? What kinds of content strategy improvements would help their organization reach their business goals? And then you will work to create practical solutions, resulting in a content strategy for that organization.
>
> **Goal:** Research an organization, assess their content, and write a content strategy for them.
>
> **Deliverables:** Three major deliverables (Pitch, Presentation, Content Strategy Document), plus a Reflection/Assessment. The format of the deliverables will vary depending on your project. You will write an

individual reflection piece describing your role on the project and your experience across the semester (1–2 pages). This kind of reflection is referred to as a post-mortem, and we'll discuss their format in class.

This document is robust, as it is meant to give students practice in several different facets of content strategy as it is practiced in industry. An important part of this assignment is the recognition that a content strategy is a living, organic document. It is meant to be updated and change as the need arises. We will briefly outline the major sections of this document; know that these sections will need to be updated on a regular basis as the role of the content strategist, the goals of the business, and the needs of the participants change over time.

- **Strategy Brief:** This section explains the purpose of the content strategy. It summarizes business goals and customer needs, and gives an overview.
- **Style Guide:** Improve the guide your team built for P2 [Project 2]. This document should be usable and useful for writers and managers in your organization. Consider multiple spaces (examples: Facebook, Pinterest, Twitter). Include Content Models (at least one) and Templates (at least 3—refined, improved from P2). Consider multiple views of one space or Website, Mobile Web, App.
- **Metadata Strategy:** How will you tag each page and item? Think through the implications for style and SEO. What keywords should they focus on?
- **Campaign Example:** Think of launching a new product, event, etc. Think of how they ought to structure the content so it could roll out with SEO considerations in place.
- **Governance Plan**: Outline the content process. Who does what, when, why? Who makes major decisions? Who needs to sign off? Include crisis communication and intellectual property/copyright.
- **Appendix:** Content Inventory, Questions/Data from interviews, testing, etc.

To be successful with teaching them how to create a content strategy, we have found it vital to work on the document during class time. The reasons are plentiful, but we find three reasons particularly important for student success. First, it gives the instructor the opportunity to walk through various themes as the semester progresses, making connections between those themes and the practice of building a content strategy. For example, a week focused on governance can help students build out a section on governance, including outlining processes and policies. Second, it gives students the opportunity to work inside the classroom with their team. Given that many students have busy lives balancing school, work, and other obligations, teaching them how to work together as a team during class can help them manage time constraints and learn how to work well together. Given that they have access to the same shared Google Doc, this team-building can help them have greater success as they collaborate online outside the classroom. Third, building the document partially during class helps students get over the fear and intimidation of its sheer size. They can see their team's progress during each class session as they learn to mimic this genre and create useful information for their clients. All this bolsters students' confidence as *padawan* content strategists.

A Note about Reflections

Both Assignment 1 (Landscape/Competitive Analysis) and Assignment 3 (Content Strategy Document) require a Reflection document to be turned in independently by the students. A Reflection is not part of Assignment 2 because this assignment is meant to roll into Assignment 3; we did not want to make this a moment to stop for big picture reflection because the teams are still in the midst of working on the larger project.

The Reflection document is a critical part of our curriculum and a central component of our assignment scaffolding. The reasons to include a Reflection are many-fold, but here we will focus on just a few reasons as to why we include them in this course. First, we want students to become reflective practitioners once they leave our institution, continuing to engage in reflexive practice (Sano-Franchini, 2017) not only when reflection documents are assigned but also as part of ethical professional practice that leads to more effective and inclusive design. Too often, decisions are made, products are launched, and processes are created with seemingly little reflection on the values embedded in and reflected by specific products. It is important that students and content strategists learn to not only reflect on the work of the team but also their part in that work. Which leads us to a second reason: self-reflection can aid in self-improvement and in the ability for students to articulate their skill set and value. This latter part is of particular importance when working to support, encourage, and prepare a diverse student population for a workplace that has historically erased the expertise of non-Western communities. The goal of the Reflection document in an Experience Architecture/Professional Writing program that values and purposely includes non-Western knowledge practices is not only to help students complete projects but also to articulate the importance of students' unique perspectives in their current and future workplaces.

Finally, reflection documents allow us to make space for students to air any issues that might be happening on the team that perhaps they did not feel comfortable communicating in another way. That said, it is so critical and important to engage with students both in and outside the classroom so that the Reflection is not simply a space to vent grievances. We would recommend reading up on group dynamics and group projects to avoid these pitfalls as you create your classroom space and as these projects progress.

Future Steps

One of the most important jobs of a technical communication instructor is to stay up to date with, if not one step ahead of, best practices, thought leadership (i.e., informed opinions and authorities in their specific fields), and technologies. This work is not easy, especially when it might feel like new ideas, concepts, and apps are being launched every day. We want to encourage you to keep current with the many professional and academic organizations that debate these practices and discuss this research. Through our own networks, conferences, and colleagues, we have learned that while there is a set of core values for content strategy, updating our syllabi, assignments, and coursework on an annual, if not biannual, basis is critical to sharing these ideas and learnings with students.

References

Andersen, R. (2014). Rhetorical work in the age of content management: Implications for the field of technical communication. *Journal of Business and Technical Communication, 28*(2), 115–157.

Batova, T., & Andersen, R. (2016). Introduction to the special issue: Content strategy-A unifying vision. *IEEE Transactions on Professional Communication, 59*(1), 2–6.

Brignull, H. (2018). Dark patterns. Retrieved from https://darkpatterns.org/.

Casey, M. (2015). *The content strategy toolkit: Methods, guidelines, and templates for getting content right.* Berkeley, CA: New Riders Press.

Gonzales, L., Potts, L., Hart-Davidson, W., & McLeod, M. (2016). Revising a content-management course for a content-strategy world. *IEEE Transactions on Professional Communication, 59*(1), 56–67.

Greenberg, S., Boring, S., Vermeulen, J., & Dostal, J. (2014). Dark patterns in proxemic interactions: A critical perspective. *Proceedings of the 2014 Conference on Designing Interactive Systems.*

Hackos, J. (2002). *Content management for dynamic web delivery.* New York, NY: John Wiley & Sons, Inc.

Halvorson, K., & Rach, M. (2012). *Content strategy for the web* (2nd ed.). Berkeley, CA: New Riders Press.

Kissane, E. (2011). *The elements of content strategy.* New York, NY: A Book Apart.

McGrane, K. (2015). *Going responsive.* New York, NY: A Book Apart.

Nielsen, J. (1994). Enhancing the explanatory power of usability heuristics, *CHI'94 Conference Proceedings.*

Potts, L., & Salvo, M. J. (Eds.). (2017). *Rhetoric and experience architecture.* West Lafayette, IN: Parlor Press.

Sano-Franchini, J. (2017). Feminist rhetorics and interaction design: A vision for facilitating socially-responsible and responsive design. In L. Potts & M. Salvo (Eds.), *Rhetoric and experience architecture* (pp. 85–108). West Lafayette, IN: Parlor Press.

Sheffield, R. (2009). *The web content strategist's bible: The complete guide to a new and lucrative career for writers of all kinds*. Atlanta, GA: Richard Sheffield.

Trice, M., & Potts, L. Building dark patterns into platforms: How GamerGate perturbed Twitter's user experience. *Present Tense: A Journal of Rhetoric in Society*, 3.6. Retrieved from www.presenttensejournal.org/volume-6/building-darkpatterns-into-platforms-how-gamergate-perturbed-twitters-user-experience/

Usability.gov. (2018). Heuristic evaluations and expert reviews. Retrieved from www.usability.gov/how-to-and-tools/methods/heuristic-evaluation.html

Wachter-Boettcher, S. (2012). *Content everywhere: Strategy and structure for future-ready content*. New York, NY: Rosenfeld Media.

Welchman, L. (2015). *Managing chaos: Digital governance by design*. Brooklyn, NY: Rosenfeld Media.

Williams, R. (2015). *The non-designer's design book: Design and typographic principles for the visual novice*. San Francisco, CA: Peachpit Press/Pearson Education.

4

TEACHING TOPIC-BASED WRITING

Yvonne Cleary

UNIVERSITY OF LIMERICK

Chapter Takeaways

- Offers an overview of the evolution of topic-based writing.
- Discusses Baker's *Every Page is Page One* (EPPO) topic characteristics.
- Proposes teaching strategies based on EPPO characteristics.
- Discusses an assignment that instructors can adapt for similar courses.

For most of the twentieth century, user assistance was developed in long-form documents, typically books and manuals. Content was "tied up in proprietary formats, in large compound documents, and focused on presentation rather than message" (Hackos, 2002, p. 1). However, from the 1980s onward, personal computers and graphical user interfaces became commonplace, and the user profile evolved to include lay users and non-IT specialists. These users needed assistance to help them negotiate the attendant software and hardware. Manuals were too long and detailed. Furthermore, emerging research demonstrated the power of minimalist, chunked information.

As a result, the past three decades have seen a shift towards short-form, topic-based content that users access for immediate queries. A respondent to a recent survey about technical communication workplace issues noted that topic-based writing "is now [the] de facto standard for all mediums" (Lanier, 2018, p. 72).

This chapter begins by outlining the evolution of topic-based writing, followed by a discussion of definitions and current implementations. It then describes Baker's (2013) *Every Page is Page One* topic-based writing characteristics. I have used Baker's book to teach topic-based writing in one technical communication graduate course. The chapter describes the topic-based writing assignment that students on that course take. I conclude by evaluating its impact and describing potential improvements and future teaching directions.

The Evolution of Topic-based Writing

John M. Carroll (1998) described the development of early mass-produced desktop computers as a "revolution in technology to support human activity" but, in the same sentence, characterized users as "terrorized" (p. 2). He explained the context in this way:

> The user population for computers was increasing and diversifying rapidly; programmers and engineers were being replaced by secretaries and professionals as the typical users. But companies did not understand the needs of these new users, and they were not prepared to technically support them.
>
> *(p. 2)*

Because of this technology "revolution," technical communicators were needed to explain new technologies. Manuals were the most common form of user assistance. They were usually long, unwieldy, and difficult to use. Carroll's seminal work on minimalism contributed to a move towards short-form, minimalist instruction (Evia, 2019). Minimalist instruction helps to reduce the learning time, cognitive load, and error rate, and, by extension, the emotional drain on users.

An important objective of minimalism is to help individuals to perform tasks quickly. Rather than describing features of a system, minimalist instruction focuses on what people can do using the system. Users are impatient and curious: they do not want to read a whole manual in order to complete a task. Extraneous information may be distracting, unhelpful, and even deleterious to their efforts. Minimalism, therefore, promotes task-oriented instruction to solve immediate problems. Its four principles (Van der Meij & Carroll, 1998, p. 21) remain bases for contemporary user assistance topics:

1. Choose an action-oriented approach.
2. Anchor the tool in the task domain.
3. Support error recognition and recovery.
4. Support reading to do, study, and locate.

According to Bellamy, Carey, and Schlotfeldt (2012, p. 5), "one of the revelations of minimalism is the task-oriented topic that focuses on the goals of the user."

Horton's 1990s textbook *Designing and Writing Online Documentation* proposed writing and design principles for online content. The book included a full chapter on writing topics. Horton also discussed how to develop online help. The help format that became prominent in software applications from the early 1990s used concepts of minimalism. Help was organized within a book structure and built around topics—individual units of information, each answering one question, or providing instructions for a unique task (Horton, 1994). Some topics within online help were conceptual, but most were procedural. Technical communicators used, and some continue to use, Help Authoring Tools like *RoboHelp* to develop online help topics.

The publication of DITA (Darwin Information Typing Architecture) in 2005 has gradually led to a paradigm shift in user assistance development. DITA is an open-source architecture for "authoring, producing, and delivering topic-oriented, information-typed content that can be reused and single-sourced in a variety of ways" (OASIS, 2018, para 1). DITA evolved from Robert Horn's (1992) Information Mapping concept by, as Mark Baker (2013) notes, "borrowing the idea that a document is a map connecting different types of content object" (p. 75). In DITA, content objects are called topics, and they are connected using DITA maps. Hackos (2016) argues that

> the introduction of the DITA standard has been of tremendous benefit to the field of technical communication. The standard supports topic-based, structured authoring in XML [extensible markup language] and the separation of form from content, ensuring that we can generate multiple output types from the same source.
>
> *(p. 29)*

Bellamy et al. (2012) agree that DITA was "born" to make topic-based writing easier (p. 5). Jason Swarts (this volume, Chapter 9) focuses on DITA in more detail. Of interest to this chapter is its topic-based orientation.

Defining Topics

Horton (1994) identified key features of topics: they are units of information, perceived by users as a unit; they answer a user's question comprehensively; and they are accessible at the level of the topic. The OASIS (2005) DITA architectural specification offers a more detailed definition, but includes features that closely mirror Horton's (1994). It defines a topic as "a unit of information with a title and content, short enough to be specific to a single subject or answer a single question, but long enough to make sense on its own and be authored as a unit." All DITA topics are structured; each piece of content is labelled, and each has a specific function.

Baker's definition sees a topic as "a small independent piece of information on a single subject" (Baker, 2013). He uses the phrase "every page is page one" to describe core principles: topics are always a starting point, cannot depend on information described elsewhere, and must serve a unique purpose (p. 71). In his recent book, Baker distinguishes between his conception of topics, as "complete rhetorical blocks," and DITA topics that often need to be combined for completeness or coherence (Baker, 2018, p. 238). Indeed, Bellamy et al.'s (2012) definition demonstrates the more restrictive nature of DITA topics: "a topic has just enough information to make sense by itself but not so much content that it covers more than one procedure, one concept, or one type of reference information" (p. 7). These differences in the interpretation of what constitutes a topic are evident in approaches to user assistance development in organizations.

Implementations of Topic-based Writing

Topic-based writing enables separation of content from structure (Hackos, 2002, 2016), using markup languages like XML or DITA. Evia (2019) explains that "[topic-based] writing is the basis of several technical communication and intelligent content practices" (p. 9), including single sourcing and component content management. Within a topic, different parts of the content need to be labelled—using metadata, for example—to make them accessible in a database or content management system (Rockley & Gollner, 2011). Baker (2013) explains that good metadata reflect the content accurately to enable filtering. As well as learning how to write topics, technical communication students also need to learn the skill of writing metadata.

Organizations with small content sets may use topic-based writing support within applications (like *FrameMaker* and *MadCap Flare*) or use XML editors (Pringle & O'Keefe, 2009). Although some software applications provide support for DITA, organizations with teams of writers and extensive content sets are more likely to use tools based on the DITA Open Toolkit, together with content management systems.

In some organizations, content continues to be created and updated using traditional word-processing and publishing tools, but topic-based writing approaches have become the most common means to develop reusable content, and DITA is the most common implementation of topic-based writing. Although DITA has been widely adopted (Andersen and Batova, 2015; Evia, Sharp, & Pérez-Quiñones, 2015), its complexity has been highlighted as a disadvantage (see, for example, Baker, 2018). Alternative, less complex, topic-based writing approaches have emerged in recent years, including Markdown and AsciiDoc. The recent publication of Lightweight (Lw)DITA aims to offer a "standards-based alternative for situations in which DITA 1.3 would be too complex" given the maturity of the DITA architecture (OASIS, 2017). LwDITA has fewer elements and a subset of the map elements, resulting in a gentler learning curve for new users and reduced maintenance requirements.

Benefits of Topic-based Writing for Organizations

Topic-based writing has become a dominant content development method because it offers several benefits for organizations, as well as users. In most workplaces, multiple writers produce information, and individual writers are assigned sets of topics to write. Bellamy et al. (2012) argue that writers are more productive in this model because they specialize. Additional benefits of topic-based writing include content reusability (Pringle & O'Keefe, 2009; Bellamy et al., 2012; Hackos, 2016). It is possible to include the same topic in more than one type of output (e.g., in a help system, website, and manual). When topic-based content is tagged, targeted to users' specific contexts, and linked to other relevant information, topics are part of a dynamic publishing model that increases opportunities for personalization and responsiveness (Hackos, 2002; Baker, 2013). Writing in topics, rather than long documents, also reduces the complexity of review processes and ensures reviewer feedback is more targeted (Bellamy et al., 2012).

Teaching Topic-based Writing

Recent technical communication textbooks have started to include sections on topic-based writing. For example, Baehr and Cargile Cook's (2017) *The Agile Communicator* includes short sections on content reuse and metadata. Teaching cases and articles about teaching aspects of topic-based writing have also begun to appear in the following sources:

- Evia et al. (2015) present a teaching case of using DITA in an undergraduate course. The layered assignment takes students through phases, from writing a standard operating procedure, to separating content from design, to XML authoring of granular content, to authoring in DITA, through to finally developing reusable content.
- Duin and Tham (2018) describe a multi-faceted assignment in which students create XML tags, and an accompanying document type definition, subsequently develop DITA topics, and use DITA maps to publish the content.
- Batova, Andersen, Evia, Sharp, and Stewart (2016) outline strategies for incorporating topic-based writing and content management into technical communication curricula. Two of the four examples they discuss do not require instructors to know XML or DITA. In one example, students conduct content audits, and in another, they develop topics in *Microsoft Word*. These rhetorically based assignments help students to learn the skills of writing topics independent of technology.

Although they are increasing in number and improving in quality, more resources are needed that outline clear topic-based or structured authoring teaching approaches and assignments that instructors can easily adapt to individual courses or programs. Furthermore, the complexity and the expense involved in adopting specialist software or a content management system make it difficult for instructors to use DITA to teach topic-based approaches to writing.

For the past three years, I have used Baker's (2013) *Every Page is Page One: Topic-Based Writing for Technical Communication and the Web* to teach principles of topic-based writing. Baker (2013) critiques the lack of rhetorical guidelines for topic-based writing in DITA. While DITA topics must include a title, a short description, and body text, further rhetorical guidance is missing. He identifies seven characteristics of so-called "every page is page one" topics. He discusses each characteristic, and provides writing advice based on these rhetorical features. These characteristics work in a teaching context both as a theoretical and practical foundation for writing topics. I have used the characteristics to teach topic-based writing in a technical communication graduate course. Below is a brief description of each characteristic, followed by a table suggesting writing guidelines and possible teaching strategies.

EPPO Characteristics

1. Topics are "self-contained". Baker distinguishes the term "self-contained" from the more common term "stand-alone" using a recipe example. A recipe is self-contained if it has all the information the reader needs: introduction, ingredients, and instructions. Although each component would be separate topics in DITA, the information is more usable in a single topic. Self-contained topics exist within a broader information context that the reader may draw upon for further information or to gain a deeper understanding. In the case of a recipe, that broader landscape is a recipe book. In the case of a knowledge base, the broader landscape is the sum of all articles about the subject.
2. Topics are specific, with a limited purpose, and the topic size is defined by its purpose. In a discussion of task- versus feature-based writing, Baker argues that, contrary to some advice, task-based writing must mention product features. Users often express their problems and search queries in terms of features. Task-based writing is distinguished from feature-based writing because the task-based topic contains "information on features that helps users accomplish tasks" (2013, p. 88), but the topic is not about the feature. Baker also argues that although every writer wants to create topics that respond precisely to an individual reader's personalized inquiry, in practice that is impossible, and topics need to serve many readers.

3. Topics "conform to a type" (Baker, 2013, p. 97). Types are effectively templates that prescribe the types of content to include. Baker provides several examples, recipes again, but also API (application programming interface) documents, user reviews, and Wikipedia entries for similar items (e.g., vehicles or cities). Indeed, DITA depends on "information typing," or categorizing content according to its function, or type. The most common DITA topic types are: Task, Concept, and Reference. Each DITA topic type has a different function, and as a result, different features.

   - Task topics tell readers how to do a procedure, usually through step-wise instructions.
   - Concept topics provide background information, definitions, or explanations, and are usually written in paragraphs.
   - Reference topics "provide quick access to facts" such as commands in a programming language, or catalogues (OASIS, 2005, para 2), and are usually in tabular format.

   These three delineations have become archetypes, even among non-DITA users. Baker (2018) is critical of the lack of nuance imposed by adhering strictly to three topic types.
4. Topics "establish their context." Regardless of how readers have arrived at a topic, they should be able to understand its function. Rhetorical devices that help to establish context include a clear title, a "succinct, context-setting first paragraph" (Baker, 2013, p. 119), or a graphic.
5. Topics "assume the reader is qualified" (p. 123) to do the task or understand the information provided. Rather than providing all possible definitions and information that a reader might need, a topic should focus on a specific purpose. It is possible to direct readers to supplementary information or definitions if necessary. Including them within every topic makes the content bloated and, ultimately, more difficult to use.
6. Topics "stay on one level" (p. 130), whether of detail, abstraction, or depth. Baker (2013) argues that although it is appropriate to write content for a book on various levels, moving from abstract to specific, theoretical to practical, or conceptual to procedural, readers use topic-based information at one level at a time. It may be helpful to provide links to other levels.
7. Topics include as many links as will be helpful, useful, or necessary. Baker (2013) notes that linking within topics is "surprisingly controversial in technical communication and content strategy" (p. 140), because it implies that the topic content is insufficient or not good enough. He argues that links are necessary because they support a user's, typically peripatetic, online behaviour. Of course, this idea is not new. Horton (1994) called links the "highways, back roads, and sidewalks of information," and he

> required them to be included in topics (p. 127). Links essentially make content accessible and findable. Manovich (2001) argued that "navigable space is a key form of new media" (p. 252) and his discussion invoked explorers and *flâneurs* to describe online trajectories and strategies.

Table 4.1 outlines some writing advice from Baker, and also proposes strategies that instructors can adapt, and resources that are helpful for teaching these strategies.

The teaching strategies can be used to teach topic-based writing, or incorporated into topic-based writing or content audit assignments, as in the assignment examples that follow.

Topic-based Writing in a Graduate Technical Communication Course

I direct a master's program in Technical Communication and E-Learning at the University of Limerick, Ireland. We also offer a Graduate Certificate program. As Program Director, I must ensure that the curriculum is fit for purpose. Cognizant of the prominence of structured and topic-based structured writing methods in technical communication workplaces, I developed a structured and topic-based writing component in one course taken by both certificate and master's students. Some students take the module online, while others come to campus.

One unit of one course focuses on topic-based writing, through lectures and invited presentations, online discussions, background readings, podcasts, and screencasts. Some additional factors have influenced my approach to teaching this content:

- Foregrounding the rhetorical skills of topic-based writing.
- Eschewing expensive tools, considering the lack of uniformity across editors and tools used in technical communication workplaces.
- Minimizing exposure to the relative complexity of DITA in particular, especially for students new to technical communication, and who are undertaking a one-year program.

The core text for this unit is Baker's (2013) *Every Page is Page One*. Baker's book is not a perfect system for teaching topic-based writing—the examples are sometimes abstract, some of the characteristics overlap, and these overlaps can be confusing for students, especially when the writing advice is very similar for different principles. Nevertheless, it provides a thorough foundation for students learning to write topics that are thoughtful, consistent, helpful, complete, and purposeful. This unit also covers the DITA standard and DITA topic types. In the past year, I have introduced LwDITA, with Carlos Evia's book *Creating Intelligent Content with Lightweight DITA* (2019) as a companion text.

TABLE 4.1 Strategies for Teaching EPPO Topics

Characteristics: Topics	*Baker's Writing Advice*	*Teaching Strategies*	*Useful Resources*
1. … are self- contained	• Avoid switching between topics when writing, and avoid thinking about the topic as part of a larger content set. • Write a brief summary of every topic at the planning stage. • Write topic metadata before writing the topic.	• Assign each student an individual topic to work on. • Review examples of task-based topics, focusing on how they are complete, but limit the scope to a unique purpose. • Identify potential metadata (e.g., author, category, or audience type); review examples of metadata; and practice writing metadata.	Chapter 14: *Every Page is Page One*, 2.1.2.3 DITA Metadata: http://docs.oasis-open.org/dita/dita/v1.3/errata02/os/complete/part3-all-inclusive/archSpec/base/dita-metadata.html#dita-metadata Dublin Core Metadata Initiative, Metadata Basics: http://dublincore.org/resources/metadata-basics/ 2.2.1.4 DITA Information Typing: http://docs.oasis-open.org/dita/dita/v1.3/errata02/os/complete/part3-all-inclusive/archSpec/base/information-typing.html#information-typing
2. … have a specific and limited purpose	• Make a list of "the obvious extensions of the current subject that this topic is not going to cover" (Baker, 2013, p. 151) as a means of keeping the content focused.	• Practice task-based writing. • Map potential user journeys through a set of topics.	
3. … conform to a type	• Examine different types of topics to identify their commonalities, and the essence of any particular type. • "Focus on what information is required to meet the specific and limited purpose of the topic type" (Baker, 2013, p. 153).	• Distinguish between tasks and procedures. • Explore the features of DITA topic types: Task, Concept, and Reference. • Discuss other potential topic types.	

4. … establish their context	• Write clear headings. • Ensure that the first paragraph is brief and clear. • Provide appropriate metadata.	• Review and discuss examples of topic headings, summaries, and short descriptions. • Identify user expectations and priorities for sample topics. • Identify related topics and links for sample topics.
5. … assume the reader is qualified	• Define what "qualified" means in terms of the content for any particular topic, by clarifying what a reader needs to know or be able to do, even if that information is not explicitly stated within the topic.	
6. … stay on one level	• Record related subjects, in the form of links or metadata. This advice overlaps somewhat with the advice for the second and seventh characteristics.	
7. … include links as necessary	• Provide a comprehensive set of links to related content, in whatever format is permitted within the authoring tool or content management system.	

Industry Input

Like in many technical communication programs, guest speakers from local companies come to campus to speak to students several times a year. These presentations are recorded for online students. This input helps to ensure that the program content is congruent with trends in practice. The presentations also enable students to recognize different ways of approaching topic-based writing, because the speakers work in various companies and team configurations, and each presentation provides a different perspective on content development practice. Most speakers use topic-based writing methods, but tools and standards differ. Some write topics using DITA, Markdown, AsciiDoc, or DocBook. Others use XML editors such as *Oxygen XML Editor*, or software such as *MadCap Flare* or *FrameMaker*, or proprietary editors. They use content management systems, including *IxiaSoft* and *SharePoint*. Some use *Microsoft Word*. The plethora of tools indicates a lack of congruence in technology. Evia et al. (2015) overcame this challenge when teaching DITA by using trial software and open-source editors.

In addition to input from local practitioners, students also learn about topic-based writing by reading practitioner blogs. For example, until 2018 Mark Barker's blog, also called *Every Page is Page One*, featured regular posts about how to write topics, and Tom Johnson writes about aspects of topic-based writing on his well-known blog *I'd Rather Be Writing*.

The Assignment

The course content is designed to prepare students to recognize rhetorical features of topics, and to undertake an individual topic-based writing assignment. The overall purpose of the assignment is to prepare students for workplaces where topic-based writing is prevalent. It also helps them to recognize LwDITA syntax, to separate the content of a topic from its appearance, and to use and recognize metadata. The assignment outline explains the value of this assignment:

> It is now common for technical communicators (and professionals in related disciplines of information development and content development) to create topics, instead of long-form documents, and to write content for re-use in many formats and outputs. Many technical communicators create content using XML-based authoring tools. This assignment also provides you with an opportunity to experiment with separating content from structure.

The assignment has evolved over the past five years. In its most recent iteration, students conduct a content audit of a topic, write a topic of up to 500 words, share tips and provide support in a peer discussion forum, and reflect on their learning.

Topic Audit

Students complete an online activity to evaluate a publicly available user assistance topic, such as a product help topic or an article published in a product knowledge base. I introduced this activity for the first time this year, following Batova et al.'s (2016) description of a content audit. This component is graded, and is conducted through the Virtual Learning Environment (VLE) discussion forum. Students respond to the following prompts:

- Identify a user assistance topic published online.
- Determine the topic type.
- Write a short post stating the topic type, and outlining whether, and in what ways, the topic incorporates Baker's characteristics of topic-based writing. Include a link to the topic in your post.

The purpose of this activity is to promote understanding, and stimulate discussion, of topic types and characteristics.

Topic-based Writing Assignment

In the second part of the assignment, students write a topic that includes the following aspects:

- Modular—content is developed at the level of a unit of information.
- Independent—each topic is complete.
- Reusable—topics can be combined in different ways to produce different outputs.

Students receive information about the topic type, audience, and proposed output format. Although students use LwDITA elements to identify different parts of topic content (including the title, short description, body text, lists, table components, and captions), they also use EPPO rhetorical characteristics. The final topic must:

1. Be self-contained.
2. Be limited in scope.
3. Respond to a specific user need.
4. Provide contextual information.
5. Conform to a type (e.g., Concept, Task, or Reference) and make that type explicit.
6. Include links to relevant additional information.

The topic must also include metadata (at least the following: author name, topic type, and audience type). Students use *Notepad++*, a free code editor, to write the topic. Students write a corresponding style sheet (using Cascading Style Sheets or XSLT) that describes the appearance of each content element. They ensure that the topic and style sheet files are linked. The output files must work in a web browser.

Learning materials include a sample topic, and a sample stylesheet. I also provide a corresponding screencast demonstrating how I created the sample files. I provide a lab sheet and screencast explaining how to use *Notepad++* to write the topic and linked stylesheet.[1]

Students receive enough guidance from the outline and materials to be able to begin problem solving, but the samples are limited in scope and they must use more elements, an original style sheet, and original content in the topic they write.

Peer Discussion and Reflection

During the assignment, students participate in online discussions to share at least one post that either asks a question, provides an answer to a question, or offers advice or a tip. The peer discussion component alleviates the discomfort of beginners and leverages the expertise within the group. Because students have different levels of experience and some have worked in topic-based writing and structured writing environments, they can, and are strongly encouraged to, support one another.

On completion, they write a forum post summarizing their experiences and reflecting on what they have learned. The reflection process promotes metacognition.

Conclusions and Teaching Directions

Topic-based writing has become an essential skill for technical communication students. Although the technical skills and applications professionals use to develop content are various and dynamic, the topic as a rhetorical unit of delivery is becoming a standard, regardless of the type of content to be developed.

This chapter has described one type of assignment in one technical communication program. The advantages of this assignment are its emphasis on the rhetorical skills involved in topic-based writing and on the separation of content and format. These correspond to core learning outcomes and to strategies students will have to deploy in the workplace. The assignment is grounded in durable principles, rather than in applications that might become obsolete or

1 Students who use iOS devices need to use an alternate application, since *Notepad++* is only available for Windows. Students tell me that they use *Atom*, *Sublime Text*, or *TextWrangler*.

irrelevant. A related advantage is the assignment's flexibility and viability. Students do not have to purchase or commit to software; the applications they use are free and widely available. It also offers opportunities for peer support and reflection on learning. This comment from an employer validates this approach:

> It is very important that students learn how to be adaptable and flexible. It is not so much about becoming an expert with a given software tool, more that students gain the learning skills that they can apply and transfer to any scenario. Every company uses different tools and these are constantly being updated and changed, depending on the project requirements. A course should teach student[s] the skills that will make them adaptable to any scenario within the workplace.

Although the assignment described in this chapter requires students to demonstrate some technical acumen, a topic-based writing task completed using a word-processing application or a wiki would work well too, and would have the advantage of enabling instructors and students to engage wholly with rhetorical principles.

At present, this is an individual assignment, albeit with collaboration opportunities. Students do not have support for combining topics to create different outputs. Furthermore, the scope to teach topic-based writing specifically, and content management more generally, is limited because this content is currently covered in one unit of a course on a busy one-year program. I plan to develop a full course on structured writing within which students will collaborate on the topic-based writing component. Students will be able to combine topics to produce different information products for different user scenarios.

Technical communication is, of course, a dynamic field. Instructors teaching this type of content must be alert and responsive. The curriculum, learning materials, and assignment described here have evolved over several years, and every year the students' experiences together with new publications and developments in the field suggest new themes, and new ways of delivering this content. Involving practitioners in a course or program helps students recognize the real-world application of their learning, and helps instructors ensure that the curriculum and course content are relevant.

References

Andersen, R., & Batova, T. (2015). The current state of component content management: An integrative literature review. *IEEE Transactions on Professional Communication*, *58*(3), 247–270B.

Baehr, C., & Cargile Cook, K. (2017). *The agile communicator: Principles and practices in technical communication* (2nd ed.). Dubuque, IA: Kendall Hunt.

Baker,M. (2011–2019). *Every page is page one: Mark baker's blog*. Retrieved from: https://everypageispageone.com/.

Baker, M. (2013). *Every page is page one: Topic-based writing for technical communication and the web*. Laguna Hills, CA: XML Press.

Baker, M. (2018). *Structured writing: rhetoric and process*. Laguna Hills, CA: XML Press.

Batova, T., Andersen, R., Evia, C., Sharp, M.R., & Stewart, J. (2016). Incorporating component content management and content strategy into technical communication curricula. *Proceedings of the 34th ACM International Conference on the Design of Communication (SIGDOC '16)*. doi: 10.1145/2987592.2987631.

Bellamy, L., Carey, M., & Schlotfeldt, J. (2012). *DITA best practices: A roadmap for writing, editing, and architecting in DITA*. Indianapolis, IN: IBM Press.

Carroll, J.M. (1998). Reconstructing minimalism. In J. M. Carroll (Ed.), *Minimalism beyond the Nurnberg funnel* (pp. 1–17). Cambridge, MA: MIT. Press.

Duin, A.H., & Tham, J.C.K. (2018). Cultivating code literacy: Course redesign through advisory board engagement. *Communication Design Quarterly*, *6*(3), 44–58.

Evia, C. (2019). *Creating intelligent content with Lightweight DITA*. New York: Routledge.

Evia, C., Sharp, M.R., & Pérez-Quiñones, M. (2015). Teaching structured authoring and DITA through rhetorical and computational thinking. *IEEE Transactions on Professional Communication*, *58*(3), 328–343.

Hackos, J.T. (2002). *Content management for dynamic web delivery*. New York: Wiley.

Hackos, J.T. (2016). International standards for information development and content management. *IEEE Transactions on Professional Communication*, *59*(1), 24–36.

Horn, R. (1992). *Developing procedures, policies and documentation version 1.01*. Waltham, MA: Information Mapping.

Horton, W. (1994). *Designing and writing online documentation: Hypermedia for self-supporting products* (2nd ed.). New York, NY: Wiley.

Johnson, T. (2006–2019). *I'd rather be writing: Exploring technical writing trends and innovations*. Retrieved from https://idratherbewriting.com/.

Lanier, C. (2018). Toward understanding important workplace issues for technical communicators. *Technical Communication*, *65*(1), 66–84.

Manovich, L. (2001). *The language of new media*. Cambridge, MA: M.I.T. Press.

OASIS (2005, May 9). Reference. Retrieved from https://docs.oasis-open.org/dita/v1.0/archspec/dita_spec_22_info_refs.html.

OASIS (2017, November 7). Lightweight DITA: An introduction version 1.0. Retrieved from http://docs.oasis-open.org/dita/LwDITA/v1.0/cnprd01/LwDITA-v1.0-cnprd01.html#why-lwdita.

OASIS (2018, June 19). 2.1 Introduction to DITA. Retrieved from http://docs.oasis-open.org/dita/dita/v1.3/errata02/os/complete/part3-all-inclusive/archSpec/base/introduction-to-dita.html#introduction-to-dita.

Pringle, A. S., & O'Keefe, S. S. (2009). *Technical writing 101: A real-world guide to planning and writing technical content*. Durham, NC: Scriptorium Publishing.

Rockley, A., & Gollner, J. (2011). An intelligent content strategy for the enterprise. *Bulletin of the American Society for Information Science and Technology*, *37*(2), 33–39.

Van der Meij, H., & Carroll, J. M. (1998). Principles and heuristics for designing minimalist instruction. In J. M. Carroll (Ed.), *Minimalism beyond the Nurnberg funnel* (pp. 19–54). Cambridge, MA: MIT Press.

5

TEACHING USABILITY STUDIES AND CONTENT MANAGEMENT IN TECHNICAL COMMUNICATION

Bill Williamson and Scott J. Kowalewski

SAGINAW VALLEY STATE UNIVERSITY

Chapter Takeaways

- Because usability studies combines user advocacy with quality assurance, it brings together both philosophical and pragmatic ends.
- As professors and professional mentors, our primary objective is to prepare students to become advocates for the stakeholder communities they will serve throughout their careers.
- By framing usability studies as *bricolage*, and by necessity the usability researcher as *bricoleur*, we construct an understanding of usability research as complex and multifaceted, requiring an assortment of problem-solving approaches.
- Scaffolded projects help students meet the challenge to think like researchers and execute like designers.

Usability studies serves as the foundation for user-centered information- and graphic-design practices. It is the methodology behind UX (user-experience design), UI (user-interface design), and XA (experience architecture), for example, and informs work in CX (customer-experience) strategies. Technical communication has evolved from creating tool- and system-centered documents toward user-centered ones. When users represent the core concern of technical communicators, their work is more likely to be empathetic, inclusive, effective, and rhetorically appropriate. Usability studies draws on a variety of disciplines, including psychology, computer science, ergonomics, graphic arts, technical communication, and more (UXPA, 2013). Two things are significant about this gathering of professional and scholarly focus: user needs and expectations are important to a diverse array of disciplines; usability studies is informed by an incredible variety of perspectives and strategies for understanding human encounters with technology and information. Such multidisciplinary insight makes usability experts more effective at serving users. In this chapter, we explore strategies for integrating usability methods into information design courses that are driven by content management systems.

The typical user of an information product cannot likely identify whether that document was designed by a single author at a computer or by a team of developers and strategists through a content management system (CMS). CMSs and single-source design methods (e.g., modular, structured content development) make certain documents (e.g., websites, user-driven knowledge bases, product catalogs, operator manuals for different models within a product line) easier to manage. However, CMSs and single-source systems are thus more meaningful as tools for engagement and design to technical communicators than they are to reader-users. Usability studies offers designers insight into user experiences for such complex and (often) dynamic information systems, making user-centered design possible.

Our chapter draws pedagogical connections among technical communication, single-source design strategies, and usability studies, first by exploring how usability studies offers insights into complex information products and design practices, and then by framing pedagogical strategies for CMS-driven courses. We describe activities for thinking and doing that can be integrated into CMS-driven writing and design courses. In our conclusion for the chapter, we briefly discuss how this examination of content management strategies and usability studies leads readily into classroom discussion of related topics, such as user-centered design, XA, professional advocacy, writing and design as *bricolage*, researcher values and responsibilities, and professional responsibility more generally.

A Brief Overview of Usability Studies

The phrase "usability studies" describes a broad spectrum of research strategies and methods that design scholars and professionals use to examine how things function. People apply this thinking in every industry, although they do not all use the same terminology to identify their work. Although several terms and phrases are applied to this work in the context of information design, among the most common are "UX," "UI," and "XA." Usability methods anticipate and observe user encounters with design artifacts and systems. As such, usability is essential to developing user-centered design strategies.

Design scholar Jakob Nielsen (2012) distills the core details of usability research in "Usability 101: Introduction to Usability." He links usability to quality, and lists five components upon which we can base an assessment of quality: *learnability*, *efficiency*, *memorability*, *errors*, and *satisfaction*. Nielsen also offers three definitions that might help us understand usability in practice:

- *Utility* (the design provides the features you need).
- *Usability* (those features are easy and pleasant to use).
- *Useful* (the design meets expectations for utility and usability).

In other words, usable designs are intuitive, logical, relatively easy to master, easy to troubleshoot, and generally easy to use. That does not mean that complex devices and processes cannot be usable. However, the more complex the design, the greater the necessity to make it logical and intuitive.

Jakob Nielsen (1993, 1994), Donald Norman (1988, 2004, 2013), and Steven Krug (2006, 2010) number among the most respected and cited usability pioneers. On their eponymous site for the Nielsen Norman Group (NNGroup.com, itself a rich resource for content related to experience design), Norman and Nielsen acknowledge that each joined the usability discussion in 1993. At least, that is when they began using terminology relevant to the area of inquiry: Norman "coined the term *user experience*" (NNGroup.com: About us) during his work at Apple Corporation, and Nielsen described *usability engineering* (1993) in the book with that title. Nielsen's (1994) heuristics for UI design (see Box 5.1) remain relevant to researchers and practitioners alike.

BOX 5.1 JACOB NIELSEN'S TEN USABILITY HEURISTICS FOR USER-INTERFACE DESIGN

Visibility of system status
Match between system and the real world
User control and freedom
Consistency and standards
Error prevention
Recognition rather than recall
Flexibility and efficiency of use
Aesthetic and minimalist design
Help users recognize, diagnose, and recover from errors
Help and documentation

Krug emerged as a leader in website design and usability during the late 1990s. His practices and publications champion early, frequent, low-stakes, commonsense evaluation of websites throughout their design and implementation stages. Krug has been an advocate for scenario-driven, think-aloud protocols, perhaps the most commonly implemented combination of methods for the study of information spaces such as websites and knowledge bases. Our pedagogy, like our research, has certainly benefited from that research strategy. We highlight these and other methods later in the chapter.

The scholarly and professional discourse on usability method and motive is rich and far-ranging in focus. A comprehensive historical record of this work is beyond the scope of this chapter. However, if you are new to the study of usability, we recommend "Technical Communication and Usability" from Ginny Redish (2010) as a beginning place. This brief retrospective on the shared history of the focal disciplines highlights key moments and contributions since the origin of usability studies in professional settings during the late 1970s. As many scholars do, Redish places the beginning of usability studies as a profession in the late 1980s or early 1990s (p. 192). Perhaps among the most valuable elements of Redish's article is her ability to identify the professionals and scholars who shaped early discussions and work strategies in usability studies. We strongly recommend this read even for people who are already working in the discipline, but who have less knowledge of usability's genesis.

User-centered Design in the Classroom: Linking Technical Communication and Usability Studies

As professors and professional mentors, our primary objective is to prepare students to become advocates for the user communities they will serve throughout their careers. To accomplish this, students must learn to anticipate and serve the needs and expectations of a variety of stakeholders connected to the organizations for whom they design information products. This in turn demands critical, data-driven, inclusive assessment of communication problems, contexts, and potential design solutions. Because usability studies combines user advocacy with quality assurance, it brings together both philosophical and pragmatic ends, and thus offers a space for shared engagement among scholars and practitioners. This is the space of user-centered design.

Given the challenge of writing content in such a way that a single source must serve in multiple information products (and thus multiple communication media and contexts), this writing challenge makes usability studies core/essential to the project of effective communication and design. Composing topics to be context-independent, consistently phrased across multiple contexts is challenging, to say the least. Effective design teams bring together content developers, content strategists, and editors just to manage the creation and assembly of content into usable information products. Without the addition of usability experts, there is less internal quality assurance, and potentially no one who bears the principal responsibility of serving as an advocate to the spectrum of stakeholders connected through any project.

Usability studies must remain conscious of the full spectrum of stakeholder needs and expectations. We must examine how documents meet expectations for search, for immediacy, for problem solving from the stakeholder's perspective (whatever that might mean), and for usefulness. Two questions might ascend from this framing of usability pedagogy in the context of single sourcing: What does it mean for us to teach usability studies in the context of information design? What does it mean for us to develop that emphasis on advocacy for stakeholders across the framework of a given project/context? To address these questions, we have argued elsewhere (Williamson & Kowalewski, 2017) for an understanding of usability studies as bricolage.

> We promote a vision of usability studies as bricolage. This is a place where once again usability and rhetoric closely parallel one another, and constructively overlap. Like James C. Raymond (1989), who poses rhetoric as bricolage, or as 'a collection of perspectives that yield useful insights in this situation or that, but always partial insights, never the whole truth' (p. 389), we enact a research toolkit that offers many tools, each with limitations, and each with useful qualities and characteristics.
>
> *(p. 40)*

By framing usability studies as bricolage, and by necessity the usability researcher as *bricoleur*, we hope to construct an understanding of usability research as complex and multifaceted, requiring an assortment of problem-solving approaches. The research toolkit, as we reference in the quote, is replete with methods for gathering data from interactions with these complex information products. We frame several of these methods below. In short, though, we want our students to understand usability research does not include a one-size-fits-all approach. Instead, we hope to prepare student-researchers to become flexible, adaptable writer-designer-researchers, who understand the complexity of information design in CMSs.

To accomplish our pedagogical goals, we approach usability studies as a methodology privileging bricolage, user-centeredness, and user advocacy. Our pedagogical practices toward usability studies, combined with our emphasis on bricolage, include a variety of usability and design research methods, driven by the particularities of any given project. The methods help fill the bricoleurs' toolkits, so they may apply the appropriate tool to help solve design challenges. We categorize these methods across three areas of design thinking:

- *Interpretive Methods* feature experts (e.g., designers, researchers, subject-matter experts) or informed users engaged in careful examination of a design. Interpretive methods are valuable for developing understanding of designs and the systems that support them.
- *Experience-Centered Methods* establish opportunities to observe users engaging with designs in clinical or authentic contexts. Experience-centered methods

are valuable for developing an understanding of how users interact with designs and the systems that support them.

- *Narrative and Dialogic Methods* provide participants opportunities either to engage in dialogue with researchers through a question and answer format, or to share their experiences engaging with/using an information product. Narrative and dialogic methods allow users to construct organic and authentic accounts of their interactions with an information product.

Like many technical communication teachers, we relied almost exclusively during our early work on scenario-driven, think-aloud protocols. This experience-centered method is built around the practice of generating authentic problem-solving scenarios that rely on the object of study (e.g., product site, owner's manual, knowledge base) for resolution. A classic video demonstration from Steve Krug (2010) allows viewers to observe a participant's attempt to determine the best plan for renting a vehicle from the information provided through a website for a rental service. During such a scenario, the researcher prompts the participant to explain throughout the engagement with the scenario the thought processes and decision-making as they unfold.

As useful and adaptable as scenario-driven protocols are, this method nonetheless represents a single tool in the bricoleur's toolkit. The challenges of meeting the needs of stakeholders across the spectrum of any individual project exceed the service capacity of any one method. Thus, we became committed to teaching and supporting a much more sophisticated array of research methods. We have integrated the methodological survey *Universal Methods of Design* (2012) by Bella Martin and Bruce Hanington as core reading for a variety of courses and research projects in usability. Martin and Hanington go well beyond think-aloud protocols, summarizing 100 methods for studying communication products and processes. Table 5.1 presents a limited sampling of methods that we have organized into the three categories we introduced earlier.

Note that some methods defy categorization using these descriptors (these methods are indicated by shaded cells in Table 5.1). For example, journey mapping spans categories, implementing strategies that can be described as either experience-centered or narrative: this method chronicles the on-going engagement of users with an object of study (e.g., app, tool) over a period of time (e.g., one week, one month). Persona development, a method that technical communicators rely upon for audience analysis, is at base an interpretive research tool. However, in the space of an on-going, comprehensive usability study, personas might be refined following observations of and discussions with participants in a CX audit or focus group (or other studies from these categories). The design bricoleur seeks opportunities to gather data (and ultimately insight) throughout the design cycle from a diverse array of stakeholders.

TABLE 5.1 Methods for Conducting Usability Studies by Category

Interpretive Usability Methods	*Experience-Centered Usability Methods*	*Dialogic and Narrative Usability Methods*
Heuristic analysis	User trail maps	Interviews
A/B comparison	CX audit	CX audit
Stakeholder maps	Critical incident technique	Focus groups
Task analysis	Design ethnography	Directed storytelling
Cognitive walk-through	Journey mapping	Journey mapping
Persona development	Card sorting	
	Scenario-driven think-aloud protocol	

Designing Courses and Curricula Dedicated to Usability Studies

As teachers and administrators of technical communication programs, our assessment and reflective practices help drive programmatic and curricular evolution. For us, usability studies has emerged as a core identity element for our department, its programs, and even its recruitment initiatives. Upon its establishment in 2010 as a separate academic unit (a process that saw Rhetoric and Professional Writing move out of the English Department), our department secured funding to construct an information research and design laboratory. That facility has since evolved to become the Center for Experience Research and Design (CERD), and serves as the hub for our Usability Research Team (URT). The URT brings together faculty, students, and alumni who engage in a variety of usability and UX design projects. The CERD features usability equipment (e.g., digital video recorders, an observation room, Techsmith *Morae*), and tools and spaces for audio and video production. We recently secured additional funding to add Makerspace capabilities, including STEM-related and web-/app-development activities. We recognize that not every program has a dedicated usability facility. However, we also recognize that not every program needs a dedicated facility to engage in meaningful usability research. In fact, the emphasis on bricolage encourages making the most out of the tools, equipment, and facilities available, even when those resources are limited.

Our Professional and Technical Writing program now features a course dedicated to usability studies as an option for completing the research-methods requirement. This Usability Studies course also serves as the centerpiece to our User Experience Design (UXD) minor. Any student might otherwise choose the course as an elective. The course has evolved to reflect our focus on methodology and bricolage. (For a more thorough discussion of usability studies in our Professional and Technical Writing major, see Kowalewski & Williamson, 2016). We note, however, that many (perhaps most) courses offered in our program include some usability-related component. UX represents a consistent commitment in all information design courses. In addition, usability methods provide core content for courses such as Instruction Writing and Design and Information Architecture, and for individual sections of Problem Solving in PTW (our introductory course) and Technical Report Writing (one of our two technical communication service courses). For our department, usability studies has become an ideal platform for blending theory and practice in audience-aware design strategies. Because these concepts are so deeply integrated, students are well able to adapt methods from multiple encounters into their work by the time they reach upper-division courses.

During the most recent offering of Usability Studies (Fall 2018, as of this writing), Kowalewski served as the instructor of record with Williamson in the role of co-instructor. The course was constructed on a group client-based service-learning project where students designed and conducted the primary research activities, analyzed the resulting data, and composed a recommendation report for their clients. Users and other integral stakeholders (e.g., clients and clients' representatives) were involved throughout the process. We emphasized the connections between usability studies and technical communication from historical and epistemological perspectives, highlighting the importance of an iterative design-research-design process, and drawing users in as participants/partners early in the design process. (See Box 5.2 for the course description, and Box 5.3 for the course objectives.)

BOX 5.2 COURSE DESCRIPTION FOR USABILITY STUDIES, FALL 2018

As user documentation and web applications become increasingly more complex and as mobile devices allow us to access information on-the-go from anywhere, technical and professional communicators have an important role to play that bridges human and technological elements.

In this course, students explore and apply advanced applications of user documentation and usability. We'll consider historical and contemporary theories of usability studies and user-centered design, while also addressing current issues and challenges facing technical and professional communicators in the areas of usability research and human-computer interaction. Utilizing a user-centered approach, emphasis is placed on research (design and methods) for resolving complex technological and communication challenges.

Course projects involve incorporating users from initial planning to deployment. We'll work closely and collaboratively with each other and clients. Along the way, we'll develop an understanding of current best practices and future considerations and apply this knowledge through course projects and discussions.

BOX 5.3 COURSE OBJECTIVES FOR USABILITY STUDIES, FALL 2018

Upon completing this course students will be able to do the following:

- Understand theoretical (cognitive and behavioral, for example) issues and challenges facing technical and professional communicators in the areas of usability, user-centered design, and human-computer interaction.
- Demonstrate proficiency with industry-standard tools and technologies employed by professional/technical communicators, who design user documentation and conduct usability research (e.g., Clearleft *Silverback* and Techsmith *Morae*).
- Apply knowledge of usability research methodologies and methods by designing a user-centered approach to solve complex and dynamic issues of human-computer (or human-document) interaction.
- Evaluate the results of usability tests and apply that information to persuade clients regarding appropriate document/interface revisions.

When he reimagined the Technical Report Writing service course around usability studies, Williamson was driven in part by feedback from alumni of the course's previous construction, and in part by on-going discussions with faculty and current students about their work in programs across campus. Although the earlier version of the course included a design analysis and a small-scale usability study of a professionally relevant tool or process, less emphasis was placed overall on those elements than there is now. Alumni signaled that the focus on usability paralleled work in their home programs on quality assurance and was therefore highly professionally relevant. Williamson's emerging perception that current students might benefit from more-plentiful opportunities to speak, write, and design from a position of expertise prompted him to raise the stakes on the design analysis project by incorporating heuristic analysis. The resulting course requires students to select a professionally relevant tool or product to study. That study integrates data from heuristic analysis (a method of design assessment that relies on artifact-specific criteria for evaluation) with data from user-centered methods of study (e.g., scenario-driven, think-aloud protocols; CX audits). The recommendations for design changes that result from this work thus represent professionally relevant expertise and are driven by data from two or more sources.

Designing Assignments for Courses Driven by Usability Studies

When designing courses for usability studies, UX, and content management, we encourage technical communication instructors to follow strategies that work for them, in terms of structure and organization. Our classes tend to feature scaffolded assignments that build upon one another toward the goal of helping students meet the challenge to think like researchers and execute like designers. Over the course of a whole-class experience, we seek to enact for students four core elements: direct engagement with stakeholders, application of multiple research methods, understanding of relationships among stakeholders, and reflection on the effectiveness of stakeholder advocacy.

The work of building a research practitioner identity (that union of scholar and designer to which we refer) begins with in-class discussions of topics such as research methods, tools, protocols, etiquette, responsibilities, and ethics, but must also include planning and executing iterative research and design cycles, establishing partnerships with stakeholders, managing and meeting client expectations, reporting project processes and outcomes, and preparing deliverables. Any methods of study that we expect students to implement during their work must receive significant attention in the time leading up to active engagement with conducting the studies themselves. Even for less-usability-intensive courses, appropriate time must be allotted to understanding how to design, execute, and process a study. These discussions pave the way for hands-on engagement with course concepts and strategies.

Course activities and objectives are predicated upon bringing together thinking and doing. The presence and participation of clients and connected stakeholders makes audience analysis (and thus any related processes such as persona development) more tangible and manageable. Face-to-face interaction with the people who depend on their research and design activities prompts students to reflect more immediately and more personally upon the significance of acting as advocates through their professional roles. Such reflection is fostered by the paced building of action and thought. Further, because we ask students to design their own studies, it is essential that we not rush through a predetermined schedule of milestones. That is not to say that we do not ask them to meet deadlines. Rather, we build in ample time for students to cycle through feelings of hesitancy and indecision, commitment to execution, and accumulating awareness of the stakes and implications of their work.

Box 5.4 presents in-class and small-scale thinking-and-doing assignments that help students build an identity for themselves around the interlinked challenges of research and design. Although some of this work (especially in-class activities) is completed collaboratively, or on display in the public forum of the classroom, these activities can also be individually immersive. Although most of these activities might feel familiar, atomic modeling is a process we developed during our work with the URT. This approach to visual modeling requires designers to reflect upon the knowledge and strategies relevant to a particular project, and to consider the needs, expectations, and core values of the project's stakeholders. (For more discussion of this research method, see Williamson & Kowalewski, 2017.)

BOX 5.4 IN-CLASS ENGAGEMENT AND SMALL-SCALE ASSIGNMENTS THAT EMPHASIZE USER-CENTERED DESIGN THINKING

Usability method demonstrations
Research tool demonstrations
Atomic modeling of designers
Persona development for stakeholders
Persona refinement based on usability study
Gathering and contemplating design heuristics
Examination of design genres relevant to project deliverables
Assessment of research methods for specific projects

Box 5.5 presents assignments that require in-depth engagement with research and design. Although we list fewer projects here, they are the logical destinations for the work of a class in usability. The key is in the details. For example, although we ask students to design and execute a usability study, we expect each designer/design team to apply multiple research methods during their study. Thus, the studies themselves are accomplished in stages, giving students time to reflect upon their assumptions, refine their approaches, and adapt in the moment to emerging (perhaps unforeseen) developments. Any design project in this context turns on whether students work with an existing document or must create their own. In either scenario, they may reflect upon the cycles of design, research, refinement, and further study.

BOX 5.5 DEEPER INVESTED ASSIGNMENTS FOR DEVELOPING USER-CENTERED DESIGN THINKING

Design and execute a usability study
Report results of individual study
Design and refine an information product

To illustrate the kind of work we require in the courses we have highlighted here, we offer details from the course and assignment descriptions relevant to Usability Studies (see Box 5.6) and to the version of Technical Report Writing offered with a usability emphasis (see Box 5.7). Although neither of these courses is taught with a specific emphasis on content management or single-sourced content, the emphasis on usability and UX translates to any course.

BOX 5.6 DESCRIPTION FROM KOWALEWSKI'S CLIENT-DRIVEN PROJECT

Overview

The purpose of this assignment is to design and conduct a user-centered, usability research project, applying the theories and methods we have discussed and utilizing the hardware and software (Camtasia, Morae, and/or Silverback) in the Center for Experience Research and Design. In small teams (no more than 3–5 people), you will determine the client and document/product for usability testing, design the usability research (including a planning outline with appropriate methods), recruit usability test participants, conduct the usability research, analyze research data, compose a recommendation report—including a highlights video, and present the project to the client and your classmates. Specifics of this project vary depending on the client/product, but the project must utilize a user-centered approach from initial conception to final report/recommendation.

Step 1
Planning Outline

The Planning Outline is the first step of this project. You'll want to discuss with the client their perspectives on the document's purpose, primary reader-users, contexts of use, and any concerns or challenges they have encountered. Use this information to begin drafting the planning outline. This outline should define the client or product, identify and discuss the problem statement (in other words, why is usability testing needed?), identify primary users and stakeholders, frame the methodology and discuss the methods that will be used to gather data (see below), foreground any potential problems or limitations, discuss team members' responsibilities, and provide a timeline for action.
**The methodology and methods section discusses the usability testing methods your team will employ to gather data. Because data triangulation is important for a comprehensive perspective, your team will need to use at least three different methods. Incorporate interpretive, experience-based, and dialogic/narrative methods. When choosing the appropriate methods for your research, consider the methods presentations we conducted in class and refer to the *Universal Methods of Design* text.

Step 2
Atomic Model Mapping

Following the rationale for atomic model mapping in Williamson and Kowalewski's "Cultivating a Rhetoric of Advocacy" article, your team will create an atomic model map that illustrates the relationships between technical communicators/usability specialists (your team), stakeholders, the client, research methods, and any other critical elements.

Step 3
Persona Profiles

As stated in the *Universal Methods of Design* text, "[p]ersonas consolidate archetypal descriptions of user behavior patterns into representative profiles ..." (132). Your team will create 3–5 personas of representative users based on interviews, discussions, and initial usability research data.

Step 4
Activity Report

Heading into the Thanksgiving break, your team will share an activity report that discusses what has been accomplished with the client project and what still needs to be completed—with a plan for completing the Recommendation Report on time. Use the framing from the Planning Outline to help guide this report.

Step 5
Recommendation Report

After your team has conducted the necessary research and analyzed the data, you will collaboratively compose the recommendation report. The report should draw include an introduction with problem statement, provide a methodology and methods section, overview the research results, discuss the research results, and offer recommendations for revisions. Your report should include visuals (charts, tables, graphs, screenshots, for example). Consider your client as the primary reader-user of this report. The final report should be 2000–2500 words, not including any appendices.

BOX 5.7 ASSIGNMENT SERIES FROM WILLIAMSON'S TECHNICAL REPORT WRITING

All three projects connect to the same core topic, which itself ties to the challenge of communicating professionally relevant knowledge. You will identify an object of study (OoS) no later than the end of week two. Because I emphasize communicating from a position of expertise in this course, you must identify an OoS with which you have expertise, and that is in some way relevant to your professional development or career objectives.

- *Design Assessment (DA).* The DA focuses on examining a professionally relevant object of study (e.g., app, device, process) with the purposes of assessing its function, and recommending upgrades to its design. The DA is challenging because it demands a balance of three kinds of vision: (1) neutral, objective examination of an object of study (through heuristic analysis); (2) disciplined, data-driven evaluation of that same OoS (through a usability study that you design and execute), and (3) experience with and insight into the design and function of the OoS.

- *Usability Study Report (USR).* The USR reports the methods, results, and conclusions of the usability study that you complete during Workshop 3 (Usability Study). For that study, you design and execute an examination of user engagement with the professional app, tool, or process that provides focus for your Design Assessment project. The report describes your method of investigation, the data gathered during that research, and your conclusions about the design. The USR is challenging because it requires you to think like an expert design researcher and attend carefully to the needs and expectations of people who might use the object of study.
- *Five Minutes of Knowledge (5MK).* The 5MK focuses on sharing professionally relevant knowledge about the object of study from your Design Assessment with an audience of peers. You design a screencast that introduces peers to your object of study, and then contextualizes that knowledge for them in their professional work. The 5MK is challenging because it positions you as an expert who has relevant knowledge to share. To do so, you must consider how your knowledge and expertise might complement and support professional peers.

Kowalewski's client project is a layered, multi-step, scaffolded project designed to offer students a flexible and constructive usability research challenge. Some of the project specifics may vary among groups depending on the particular client, but we work closely with students to ensure that each project is manageable within the duration of eight weeks (the project runs the second half of a 15-week semester) and appropriate for the scope of the project. Each team completes common steps, as outlined in Box 5.6, and triangulates data through a mixed-method research design. Perhaps most important, student-researchers consider the implications of the work for multiple stakeholders, including primary users, from the initial stage of research on. For example, the planning outline requires teams to have had initial conversations with their clients and other stakeholders.

Beyond the planning outline, student-researchers also complete atomic model maps and persona profiles. The atomic model mapping (again, see Williamson & Kowalewski, 2017) seeks to identify the dynamic relationships that exist in any given UX situation. We engage in this mapping process primarily as "a heuristic tool for examining professional roles and actions. Positioned in this way, the technical communicator has both the power *and* the responsibility to serve the best interests of the stakeholders" (Williamson & Kowalewski, 2017, p. 43). Atomic model maps help teams visualize relationships between researchers and clients, clients and users, researchers and users, and research methods and theoretical framings. Similarly, persona profiles allow research teams to visualize and better understand the nuances of user needs, expectations, and motivations.

Whereas the planning outline, atomic model mapping, and persona profiles are designed to be formative, the activity report is an opportunity for teams to reflect on their progress before the final project is due. Teams measure their progress in part based on the planning outline. Although we recognize that projects may veer from their original proposals, the planning outline offers a good starting point to gauge progress. As instructors, we use this reflection to offer feedback on progress and suggestions for timely completion. The project culminates in the recommendation report, where teams present their research results and provide their suggestions for document/interface/space revisions.

This project could certainly be modified to fit other contexts or courses. Within single-source and CMS environments, research teams might explore multiple output modes across user groups, for example. We encourage other teachers to create a version of this project that works for them, but we also encourage those instructors to think about the project in steps (or layers) that allow students to plan, research, design, and reflect.

We move from a project in a dedicated usability course to a project situated in the context of a technical communication service course that includes a usability focus. Technical Report Writing draws enrollment primarily from our College of Science, Engineering, and Technology (most significantly from programs in Engineering and Computer Science). Williamson's version of the course engages students through three assignments that revolve around a single professionally relevant object of study: the Design Assessment, the Usability Study Report, and the 5 Minutes of Knowledge screencast (see Box 5.7).

The Design Assessment and Usability Study Report projects require students to apply usability research methods to the study of a professionally relevant design. Students are offered significant latitude in selecting an object of study for the assignment sequence (as is illustrated in Box 5.8).

BOX 5.8 SELECTING AN OBJECT OF STUDY FOR THE TECHNICAL REPORT WRITING PROJECT SERIES

- *Select a design with which you have a history and connection*
- To speak credibly about a design, you must understand it thoroughly. The only way to accomplish that state is to experience what it means to use the thing itself. With that in mind, select your object of study based on its professional relevance and on your familiarity with it.

Remember, however, that you must also select an object of study that is in some way professionally relevant. That means it must represent an area of expertise that you are in the process of developing or an industry within which you might seek employment, or a product category significant to your career goals. It might be something you use to do your job or something that results from your job.

Begin by listing for yourself at least three or four potential foci for the project. Consider the following possibilities:

- Information Products. Books, magazines, periodicals, manuals, advertising, websites, apps, software, tutorials, animations, documents, services, and so on.
- Physical Designs. Household objects, public or private spaces, buildings, signs, and so on.
- Processes. Instructions for completing some task itself, and so on.
- Other designs/Processes. Television programming/film, music/musical genre, game/game board, professional habits, intellectual processes, and so on.

Your DA can focus on an object of study from any category. To complete the assignment, you must be able to contextualize your assessment of the design and recommend specific changes that will address its flaws. Keep that in mind while selecting an object of study.

Select a specific design example rather than a general category of things. That is, examine the January issue of *Wired*, rather than all magazines. Examine a 1968 Ford Custom 500, rather than all cars. Examine a ride on the Raptor at Cedar Point, rather than riding roller coasters in general.

The assignment sequence depends on the application of interpretive usability methods (specifically heuristic analysis during the Design Assessment project) and experience-centered usability methods (during the Usability Study that leads to the Usability Study Report). Students locate heuristic criteria applicable to their objects of study from professional or scholarly sources. For example, students who study a design that includes an interface of some sort might apply Nielsen's ten usability heuristics for UI design that we shared in Box 5.1. To complete the Usability Study Report, students design small-scale studies that involve three to five users from an appropriate user group as participants. These studies combine the think-aloud protocol method with a second method (e.g., critical incident technique, CX audit, scenario-driven encounter). Although the Usability Study Report is a stand-alone project that requires its own report submission, students must also distill that report into their final Design Assessment. The culmination of their research is a set of recommendations for improving the design of the object of study, complete with justification for how each change would make the object of study more usable in its context of use. This sequencing is presented visually in Figure 5.1.

The three projects together require students to identify areas of expertise for themselves, select an object of study that is somehow relevant to their expertise and professional development goals, and then examine that object of study based on their experience, knowledge, and original research. By design (if not always by practice), their own reflection and research regarding the object of study moves students toward greater understanding of their research subjects and the people who encounter those tools and processes. By the time they present their objects of study through the 5 Minutes of Knowledge screencast, they should be able to speak with even greater authority about their work and its focus.

The highlight assignments we present here offer complementary approaches to engaging students in the work of usability studies. This kind of work is complex and challenging, and requires the instructor to provide a supportive, exploratory environment, especially for students who have little or no experience with designing and executing experimental studies. However, by moving from in-class demonstrations and practice runs to assignments where students must claim and enact their own professional authority, it is our experience that

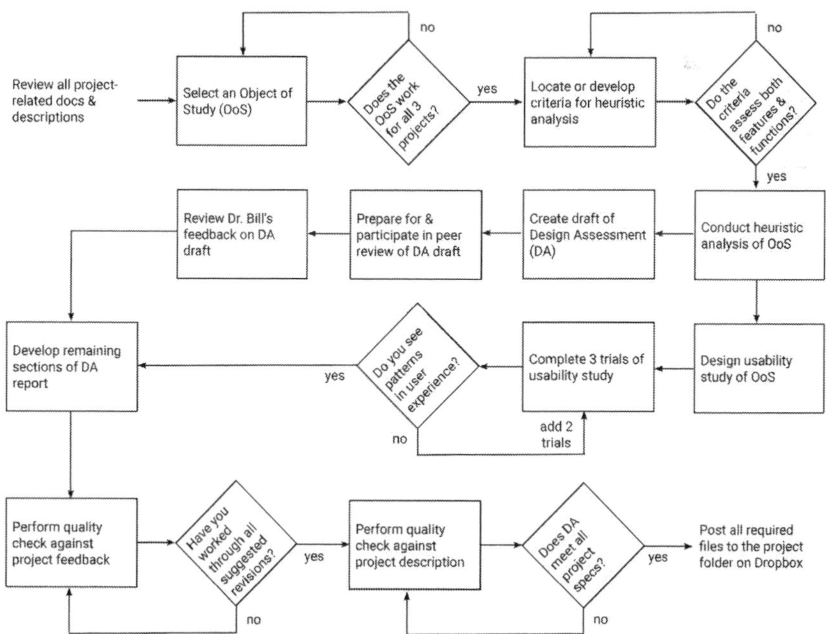

FIGURE 5.1 Process for completing the Design Assessment Project

usability-driven work fosters professional development and confidence. Any of these assignments and accompanying work processes can be integrated into a course with a content management connection.

Some Final Thoughts on the Pedagogical Relationship between CMS and UX/XA

Information design is a complex endeavor. Creating documents that communicate to readers (or users, or visitors, or however we might conceive of our audiences) is a complex process. Usability studies—as knowledge area, mindset, and methodology—offers a foundation for a user-centered research and design process that connects technical communicators critically and reflectively with their audiences and stakeholders. As structured authoring and content management strategies continue to become prevalent and more complex for technical communication faculty, researchers, and practitioners, usability and UX within these contexts must match that prevalence and complexity. Failing to value the role of usability and UX research in the structured authoring process risks creating documentation that does not meet users' utility, usability, or usefulness needs—essentially ignoring decades of academic and practitioner arguments toward user-centeredness.

In this chapter, we set out to provide actionable approaches to integrating usability and UX research practices into complex CMS and structured authoring contexts. Framing usability specialists, and by extension technical communicators, as bricoleurs has allowed us to help our students become adaptable, flexible problem solvers. Carefully layered projects, such as the usability client project and the object of study series, have provided our students with opportunities to engage with contemporary, relevant, context-specific design challenges.

We certainly do not presume our model to be the only way to incorporate usability/UX into CMS and structured authoring contexts, but we offer our journey and evolution here to illustrate one viable and sustainable approach to integrating usability and UX practices into technical writing curricula. In closing, we offer four elements we believe are key to any course/curriculum focused on usability research, especially those curricula including structured authoring and CMS.

Usability as Research Methodology

Our approach to usability studies within our curriculum has evolved from a focus on usability as a single method (think-aloud protocol) to a more robust understanding of usability as a research methodology that encompasses a variety of research methods. Framing usability studies as a research methodology helps students position themselves as problem-solving researchers who critically engage information products through an iterative process, utilizing appropriate tools to find applicable solutions.

Design Awareness

Any curriculum or course that includes usability studies should also provide students with a strong foundation in design awareness. Robin Williams' (2014) principles of design are a great starting point, but we often include Donald Norman's *The Design of Everyday Things* (2013) to help students recognize design elements such as *affordances*, *constraints*, *mapping*, and *feedback*, and to develop awareness of the human dimensions of effective design. These elements may be of particular importance in CMS environments, where writers may not have complete control over the final deliverables.

Rhetorical Awareness

Rhetorical awareness provides the underlying humanistic, scholarly foundation upon which technical communication has evolved over the past four decades. In the context of a usability course, rhetorical awareness manifests through a focus on user-centered design. Documents and interfaces are created for people, which makes them inherently rhetorical. In our courses, we draw on the work of scholars such as Robert Johnson (1997) and Barbara Mirel (2002) to help our students explore the implications of user-centered design.

Client-driven Projects

Finally, we strongly recommend that usability curriculum to include client-driven activities, when appropriate. As the body of literature surrounding service learning and client-driven pedagogy in technical communication often suggests, projects that require students to work with external clients often best embody the challenges of working in professional contexts. We recommend scaffolding these projects (see Box 5.4) to help students manage their time and resources. Students should also have ample opportunity throughout the work process to reflect, discuss, and receive guiding feedback. Client-driven projects gather the first three elements of this list in ways students find tangible and meaningful. In structured authoring and CMS environments, client-driven projects create opportunities to analyze both the process and the product of an information or interface artifact.

We conclude this chapter with a list of resources we have found valuable for integrating usability studies and UX design into our classes and curriculum. We hope readers will find this resource just as valuable and encourage them to add to this list as they discover their own useful resources.

BOX 5.9 USEFUL RESOURCES FOR USABILITY AND UX DESIGN

User Experience Professionals Association (UXPA)—USPA.org/
Professional organization for usability and UX practitioners. The site includes a code of conduct and UX definitions.

Usability.gov—Usability.gov/
A US Government site with information for implementing usability into project management processes.

Section 508.gov—Section508.gov/
A US Government site that offers resources for designing accessible products and interfaces.

Nielsen Norman Group—NNGroup.com/
The collective organization of usability and design leaders Jakob Nielsen and Donald Norman, offering web articles, training, and consulting.

UX Magazine—UXMag.com/
Digital publication for the professional organization Design for Experience.

World Usability Day—WorldUsabilityDay.org/
An annual celebration with events around the world raising awareness for usability in our everyday lives.

References

Johnson, R. (1997). Audience involved: Toward a participatory model of writing. *Computers and Composition, 14*, 361–376.

Kowalewski, S.J., & Williamson, B. (2016). Strategic assessment and usability studies: Tracing the evolution of identity and community engagement in an undergraduate professional and technical writing program. *Programmatic Perspectives, 8*(2), 96–118.

Krug, S. (2006). *Don't make me think: A common sense approach to web usability* (2nd ed.). Berkeley, CA: New Riders.

Krug, S. (2010). *Rocket surgery made easy: The do-it-yourself guide to finding and fixing usability problems*. Berkeley, CA: New Riders.

Martin, B., & Hanington, B. (2012). *Universal methods of design: 100 ways to research complex problems, develop innovative ideas, and design effective solutions*. Beverly, MA: Rockport Publishers.

Mirel, B. (2002). Advancing a vision of usability. In B. Mirel & R. Spilka (Eds.), *Reshaping technical communication* (pp. 165–188). Mahwah, NJ: Earlbaum.

Nielsen, J. (1993). *Usability engineering*. San Diego, CA: Academic Press.

Nielsen, J. (1994). Heuristic evaluation. In J. Nielsen & R. L. Mack (Eds.), *Usability inspection methods* (pp. 25–62). New York, NY: John Wiley & Sons.

Nielsen, J. (2012). Usability 101: Introduction to usability. Nielsen Norman Group. Retrieved from www.nngroup.com/articles/usability-101-introduction-to-usability/.

Norman, D. (1988). *The psychology of everyday things*. New York, NY: Basic Books.

Norman, D. (2004). *Emotional design: Why we love (or hate) everyday things*. New York, NY: Basic Books.

Norman, D. (2013). *The design of everything things: Revised and expanded edition*. New York, NY: Basic Books.

Raymond, J.C. (1989). Rhetoric as bricolage: Theory and its limits in legal and other sorts of discourse. In C.B. Matalene (Ed.), *Worlds of Writing: Teaching and learning in discourse communities of work* (pp. 388–399). New York: Random House.

Redish, J.C. (2010, September). Technical communication and usability: Intertwined strands and mutual influences. *IEEE Transactions on Professional Communication, 53*(3), 191–201.

User-Experience Professionals Association. (2013). About UX. Retrieved September 7, 2019, from http://UXPA.org/resources/about-ux.

Williams, R. (2014). *Nondesigner's design book*. San Francisco, CA: Peachpit Press.

Williamson, B., & Kowalewski, S.J. (November 2017). Cultivating a rhetoric of advocacy for usability studies and user-centered design. *Communication Design Quarterly, 5*(3), 35–47.

6

TEACHING CONTENT MANAGEMENT WITH XML

Becky Jo Gesteland

WEBER STATE UNIVERSITY

Chapter Takeaways

- Shows how to teach basic Extensible Markup Language (XML) in a one-semester, undergraduate content management course.
- Outlines the reading materials and responses in the course.
- Describes two assignments in detail (a HTML and CSS website, and XML document).
- Reinforces the value of hand-coding for students' career readiness and technical acuity.

Coming to XML

On February 10, 1998, the World Wide Web Consortium (W3) released the recommended XML 1.0 standard. Twenty-one years later, I continue to teach a brief section about XML in my Content Management (CM) course—a course required in the undergraduate Professional & Technical Writing (P&TW) program at Weber State University. This chapter explains my rationale for teaching XML, outlines the class readings and assignments, and describes two assignments in detail: the HTML and CSS Website and the XML Document.

Prior to the publication of my last chapter about teaching XML (McShane, 2009), I had taught our program's new course, ENGL 4110 Content Management, only once. Since then, I have taught the course every spring, and have adjusted the readings, assignments, discussions, tools, and so on, as the CM landscape shifts. In that chapter, I argue that we should teach XML not only because of its extensibility, or "text-appeal," but because I believe it to be "a logical place for technical communicators to locate themselves as experts" (McShane, 2009, p. 83). Moreover, I argue that "we must teach it [XML] with the theory (single sourcing), the methodology (modular writing), and the technology (content management) to support, apply, and guide it" (p. 83). Based on employer demand and changing workplace practices our program added the CM course over 12 years ago (see Thomas & McShane, 2007). Meanwhile, the field has continued to evolve, at a rapid and accelerating pace; therefore, I find myself leaning toward a nimbler approach to teaching the technology of CM.

In "Rationalizing and Rhetoricizing Content Management," Pullman and Gu's introduction to the 2008 special issue of *Technical Communication Quarterly* ("Content Management and Technical Communication"), the authors outline their concern with the lack of involvement of technical communicators in content management system (CMS) design and implementation. Sharing their concern—and as a way to address that concern, at least in part—I determined that students needed to learn something about the CM process. One approach is XML, which has the dual advantage of introducing students to the CMS process while also teaching them a bit about coding. In this special issue, Robidoux (2008) calls for students to learn "how to write and use structured writing" and delineates a comprehensive single-sourcing curriculum that includes four modules: (1) Defining structure, (2) Structuring content, (3) Analyzing content, and (4) Reusing content (p. 111). Although compelling, I find Robidoux's approach overwhelming for a one-semester class and the material too dense for students often unfamiliar with technical writing, let alone structured content.

Recent articles about teaching XML include the more accessible methods described by Batova, Andersen, Evia, Sharp, and Stewart (2016) in "Incorporating Component Content Management and Content Strategy into Technical Communication Curricula," which outlines four ways to teach content strategy in existing courses, and Robles and Frith's (2018) "Developing a Content Strategy Course and Interdisciplinary Skills: A Teaching Case," which suggests teaching XML but doing so in a separate course *or* teaching it in a content strategy course "in an assignment unrelated to the client." This latter suggestion corresponds to my approach. Whether students undertake a project that involves XML (some actually do pursue coding with their community partners), the stand-alone assignment ensures that they all have experience with XML.

Quest for a Text

The CM course fulfills the fifth learning outcome for the P&TW Program:

> *Students should construct documentation projects using single-sourcing and modular-writing principles.*

The course includes the following objectives:

> In this course, you will learn how to manage content; you will learn how to divide content into smaller chunks and re-configure those chunks into usable structures. Along the way you will encounter a variety of terms: single sourcing, modular writing, structured authoring, and information architecture. You will apply these terms and incorporate these methodologies and principles as you construct documentation projects. Moreover, you will learn how to use XML and other open-source tools.

In addition, by the end of the class students will be able to

- Use single-sourcing principles to write structured documents.
- Map content for reuse.
- Evaluate and utilize available authoring tools.
- Develop project management strategies.

Based on these outcomes and recognizing the ever-changing landscape of CM technology, I struggled to find applicable texts. I have adopted various books (Castro, 2001; Ethier & Houser, 2001; Ament, 2002; Hackos 2002a; Rockley, 2002; Glushko & McGrath, 2005; Cowan, 2008; Applen & McDaniel, 2009) throughout the years, but none suited my course objectives. In 2003, I was encouraged by the publication of Pringle and O'Keefe's *Technical Writing 101: A Real-World Guide to Planning and Writing Technical Documentation*, from Scriptorium Press. The text included two chapters: one on single sourcing and one titled "Structured Authoring with XML—the Next Big Thing." Still I could not justify requiring students to buy the text for only two chapters. I determined that the best way to provide the necessary theoretical background for CM was to survey the field through my own curated collection of readings.

Sequence of Topics and Assignments + XML

I structure the course around selected readings, associated responses, and various technical and project-based assignments. In the following sections, I outline the course while describing my reasons for choosing particular readings and my methods for scaffolding the technical assignments that culminate with an XML document. For a complete list of topics, readings, and prompts, see Table 6.1.

Coding

Most students in our P&TW program are English or Creative Writing majors who choose a minor in P&TW because they are worried about finding a job after they graduate. Thus, many of them arrive to the class full of trepidation,

TABLE 6.1 Weekly Reading Responses

Week	*Topic*	*Readings and Prompts*
1	Coding	Hipps, "To Write Better Code, Read Virginia Woolf" Do you agree with the argument? Why, or why not? What are your attitudes toward coding and technology in general? How comfortable are you with new software?
2	Single Sourcing [SS]	Rockley, "The Impact of Single Sourcing and Technology" How does she define the phrase? What impact did she believe single sourcing would have? Were her predictions correct? Why, or why not?
3	Community Partners	Which community partner do you want to work with? Why? What interests you about the organization? Please discuss a little bit and rank the organizations.
4	Content/ Presentation	Clark, "Content Management and the Separation of Presentation and Content" How do content and presentation work together? Why might you want to separate the content from the presentation? In what ways does HTML help/hinder such a separation?
5	CM & SS	Hart-Davidson, "Content Management: Beyond Single Sourcing" How does the article complicate, enlarge, and/or frustrate your understanding of content management and single sourcing?
6	XML	McShane, "Why We Should Teach XML" What is the value of XML as a coding language for content management?
7	CM Project	Lewis, "Learning to Manage Your Content" How does this conceptualization of content help you think about your CM project?
8	CM & the Workplace	McCarthy et al., "Content Management in the Workplace" What are the results of their study and how do their conclusions apply (or not) to your technical writing career?
9	The Age of CM	Andersen, "Rhetorical Work in the Age of CM: Implications for the Field of Technical Communication" What are the potential effects of her survey on the field of technical communication?
10	CM Wrap-up	1. Define "content management" What are some key concepts that you associate with content management (CM)? When would you want to apply these concepts? How might you use CM in the future? 2. Batova & Anderson, "A Systematic Literature Review of Changes in Roles/Skills in Component Content Management Environments and Implications for Education" What do you think of the term "component content management"? Does is more accurately capture the concepts we've been learning about this semester? Why, or why not?

and so I begin with a nontechnical article about the value of combining humanities and technology. Hipps' (2016) brief article in the *New York Times*, "To Write Better Code, Read Virginia Woolf," provides a nice introduction to the process of coding and encourages students to feel confident about their "soft" skill set.

After this initial reading and response, students create a Software Skills Summary, which serves as an inventory and rating system of their technical expertise. For this assignment, students begin by listing every type of software they have ever used (word processing, presentation, spreadsheet, web, email, photo, drawing, database, applications, and so on). They then explain in a sentence or two how they have used each tool. Finally, they rate their proficiency with each tool according to the "Stages of Use" model developed by Dreyfus and Dreyfus in *Mind over Machine*—for simplicity, I use Hackos' (2008) adaptation of the model, "When is Minimalism the Best Course?" The model consists of the following stages:

- Stage 1: Novice.
- Stage 2: Advanced Beginner.
- Stage 3: Competent Performer.
- Stage 4: Proficient Performer.
- Stage 5: Expert Performer.

The inventory and rating process allows students to realize how many tools they use or have used and to discover how proficient they are—even as they discover gaps in their knowledge.

Single Sourcing

The following week we read Ann Rockley's 2001 article, "The Impact of Single Sourcing and Technology," which offers one of the earliest articulations of the paradigm shift. This brief article traces the history of desktop publishing: from the 1980s, when desktop publishing allowed technical writers to do it all (page layout, document design, writing, editing, and so on) to the 2000s, when single sourcing allows us to write information once and use it over and over again. Rockley argues that single sourcing has forever changed the way we communicate. We have moved to object-oriented information, which requires structured information models and writing for reuse. Therefore, she maintains, we need to separate input (content) from output (media or information type). Furthermore, technical writing must shift to a team approach. I confess that students are often baffled by this article. How has writing changed so drastically? As we work through the subsequent readings and assignments, I bring in examples from technology they frequently use (newsfeeds and apps) and explain how these tools rely on the concept of structured content to relay information.

Community Partners

If students choose the community-engaged learning (CEL) option, they and their team will work with a community partner to develop a CM project. The CEL component requires at least 20 hours of "service," which includes all meetings with their team and community partner, any research associated with the CM project, and the time spent developing the project. Prior to the beginning of the semester, I pre-select several community partners who have need of some CM project. Past projects have included a social media platform, a college website, a donor database, and a set of policies and procedures.

HTML and CSS Website

While determining the type of CM project they will undertake, I introduce students to the concept of coding: they create a simple one-page website using a text editor and some basic styling. This introductory coding culminates in two files (.html and .css) that they upload to a Weber State University server. I use W3Schools tutorials (www.w3schools.com/), and we work through the basic tutorials together in class—I always teach this class in a computer classroom. Although they have access to *Dreamweaver*, I discourage students from using tools other than simple text editors, such as *Notepad* or *TextWrangler*. I demonstrate how to create the HTML and CSS documents and how to upload files to the server using *TextWrangler* and *FileZilla*. The assignment description includes the following information:

> For this assignment, you'll create a simple website using HTML and CSS.
>
> 1. Complete the W3Schools HTML Tutorial and CSS Tutorial.
> 2. Use TextWrangler or Notepad to create your content.
> 3. Upload your content to the Weber server using either Text Wrangler (Mac) or FileZilla (PC).
>
> To begin, you'll want to designate a home page, such as index.html. The URL for your website consists of http://student.weber.edu/username, where username is your WSU username. For instance, my website would be http://student.weber.edu/bgesteland/index.html
>
> When you're ready, you'll upload your content to the Weber server using either Text Wrangler or FileZilla—see the FTPS Filezilla Tutorial. NOTE: Due to WSU's firewall, you may need to be on campus to upload files to the server. Enter the following when prompted:

Host: student.weber.edu
Username: ad\yourWSUusername – the ad\ is important
Password: your eWeber password

I'll use the following criteria to evaluate your website:

Website		
Criteria	*Ratings*	*Pts*
Does the home page orient the user to the site?		20.0 pts
Does the site use a CSS or other design features consistently?		20.0 pts
Is the site logically organized?		20.0 pts
Do the links work?		20.0 pts
Is the site easy to use?		20.0 pts
		Total Points: 100.00 pts

The creation of HTML and CSS files requires significant time and close attention to detail; however, I have discovered that the hands-on experience instills confidence as students see their code transform into live websites.

Content/Presentation

Students then read "Content Management and the Separation of Presentation and Content," in which Clark (2008) explains that the rapid reuse and repurposing of content necessitates the separation of presentation (lexis) from content (logos). Clark argues that this separation is the foundation of CM, and that CM *systems* are the tools used to separate the presentation from content. Clark defines *content* as "a thing to be created, stored, and managed," and *presentation* as "a thing to be added just in time for the content to appear in a form suitable for human use" (2008, p. 40). An example of content is XML; examples of presentation are HTML and CSS. At this point, after working with HTML and applying styles with CSS, students understand the concept of content/presentation separation. That said, as Clark explains, it is impossible to entirely separate the two—there is always some kind of interface, even if it is a plain text editor.

Content Management and Single Sourcing

Our next reading, Hart-Davidson's (2010) "Content Management: Beyond Single Sourcing," defines *content management* as "a set of practices for handling information, including how it is created, stored, retrieved, formatted, and styled for delivery" (p. 130). He explains that the goals of CM are to distribute authoring tasks, enable multiple-audience adaptation, permit multiple output formats, and facilitate systematic reuse (p. 130). Furthermore, Hart-Davidson argues that there is value in creating content that is reusable, that technical writers *will be* content managers, and that they will need to take on varied organizational roles and responsibilities. That said, he notes the risks of CM: outsourcing and work fragmentation. At this point in the semester, this article helps deepen students' understanding of the evolving field of CM and increase their appreciation for technical skills.

XML Document

Given that students have already created a website using hand-coded HTML and CSS, the XML comes fairly easily. Most get excited about the prospect of writing their own tags and making sure the code works when displayed in a browser. I assign O'Keefe's (2009) "The ABCs of XML" because students appreciate its brevity (two pages) and its simplicity (the ABCs). Again, I use W3Schools tutorials and work through the basic tutorials together in class. I discourage students from using tools other than simple text editors, such as *Notepad* or *TextWrangler*. As with the previous HTML + CSS assignment, I demonstrate how to create the XML document and its associated CSS file and how to upload the files to the server using *TextWrangler* and *FileZilla*. Once the files are available, I show students how to create a link to their XML document from their HTML page. The assignment description includes the following information:

> For this assignment, you will create an XML document that includes a cascading style sheet (CSS).
>
> 1. Complete the W3Schools "XML Tutorial."
> 2. Use Notepad or TextWrangler to create two separate documents.
> 3. Upload your XML document and CSS to the student.weber.edu server.
> 4. Link the XML document to your HTML file.
>
> See examples below:
>
> **XML Document** (save as a .xml file)

```
<?xml version="1.0"?>
<?xml-stylesheet type="text/css" href="poem.css"?>
<poem>
<stanza1>
<line1>My love is like a summer's day</line1>
<line2>when you're gone I get so lonely I could die</line2>
```

```
<line3>so please don't leave again!</line3>
</stanza1>
<stanza2>
<line1>Your love is like a summer's day</line1>
<line2>when I'm gone you get so lonely you could die</line2>
<line3>so please don't leave again!</line3>
</stanza2>
</poem>
```

CSS Document (save as a .css file)

```
poem
{color: blue; font-family: Verdana, sans-serif; background: #CCFF00;}
stanza1
{display: block; padding:25px; text-align:left}
stanza2
{display: block; padding:25px; text-align:left}
line1
{display: block; text-align:left}
line2
{display: block; text-align:left}
line3
{display: block; text-align:left}
```

I'll use the following criteria to evaluate your XML document:

XML Document		
Criteria	*Ratings*	*Pts*
Are the tags properly nested?		25.0 pts
Does it reference a CSS?		25.0 pts
Is the XML document valid?		25.0 pts
Does it display properly in a browser?		25.0 pts
		Total Points: 100.00 pts

Hand-coding XML involves significant detailed work: checking tags, verifying consistent capitalization, and so on. But students exclaim with delight when their XML documents appear on-screen—error-free and valid.

XML

In the introduction to their edited collection (2008), Pullman and Gu define CM as "a series of regulated steps taken by an organization to ensure control and integrity of information as it goes from creation to dissemination" (p. 2). This concept resonates now because students have learned how to chunk and label content for reuse using XML. After creating their own XML documents, I have students read my article, "Why We *Should* Teach XML" (McShane, 2009). Here I explained my prior approach to teaching XML: students worked through a series of instructions, and then I quizzed them on their knowledge. Each quiz tested a step in the process of readying an XML document for use. Now, rather than quizzes, students create their own XML documents. And now, rather than providing a set of instructions, we have easy-to-use tutorials online. Nonetheless, the chapter traces my original inspiration for teaching XML and provides helpful context for students. Plus, I believe that my final argument remains relevant: "Most scholars agree that we should teach content management, single sourcing, XML, and other relatively new approaches to technical communication, but they caution that we must teach these concepts critically" (p. 83).

Content Management Project

In his short article "Learning to Manage Your Content," Lewis (2012) draws a parallel between learning to ride a bike and managing content: both are really difficult at first, so you need to break down the problem into its components and learn one new skill at a time. He outlines a simple process: perform a content audit, create a reuse map, and develop information models. I find his approach an easy way to introduce students to the next portion of the class: the CM Project. The remainder of the class focuses on this project—the remaining assignments serve as scaffolding—so we begin with a CM Proposal, in which students propose a CM project that structures content into usable chunks, labels content for reuse, and/or organizes content into containers. The project discussed throughout this section involves several steps:

1. Decide on a topic that could benefit from content reuse (Proposal).
2. Evaluate tools appropriate for your project (Tools Evaluation).
3. Create an information model.
4. Develop the content using the tool(s) of choice.

The first step is writing a proposal, which should:

- Describe in detail the situation that warrants CM. (Problem)
- Explain how they hope to manage the content. (Objective)
- Provide the specific steps they will take to accomplish their objective. (Tasks)
- Detail when they will accomplish each task. (Timeline)

Although students often complain that they do not have all the information they need to write a complete proposal, I assure them that every project plan changes and that at least they will have deadlines toward which to work.

Content Management and the Workplace

Because they now have a topic for their project and have begun content creation and/or manipulation, students are ready for a larger context, so I have them read McCarthy et al., "CM in the Workplace: Community, Context and a New Way to Organize Writing"—a study of workplace writing, pre- and post-implementation of a CMS (2011). The results of this implementation are the systematization of workflows; the introduction of new genres; the innovative use of nondigital tools; a reliance on metadata; the encouragement of changes in authorship, sharing, and reuse; and a focus on CM rather than presentation. In other words, many workplace changes are rather daunting for students to contemplate, but the article reinforces the value of CM for future career success.

At this point, I assign the Tools Evaluation (Step 2), in which students evaluate potential software for their CM project. This assignment involves two parts: developing criteria for their CM projects and researching the most appropriate tools for their projects. Recognizing that they will not have access to most CM tools—the computer classroom is limited to the Adobe Creative Suite and *Microsoft Office*—I encourage them to investigate *all* tools and perhaps experiment with some "demo" versions. After selecting a handful of tools, they analyze and evaluate the tools according to the criteria they have established. The evaluative process shows students the array of options available and teaches them how to advocate for tool adoption.

The Age of Content Management

Our next reading is Andersen's (2014) "Rhetorical Work in the Age of CM: Implications for the Field of Technical Communication," which affirms that the technical communication field has moved from a document-based to a topic-based approach to content. Technical communicators now must focus on processes, methodologies, and technologies. And the ideal content is "intelligent, nimble, smart, portable, and future-ready" (p. 116). In our topic-based world, the document is dead (p. 125). For students who took my first CM class in 2007, this proclamation would have caused a revolt. Now, more than a decade later, they get it. We build documents from chunks of information that we label and compile for various purposes, audiences, and contexts. They are ready for the information model (Step 3), which describes all the content for their CM project.

To better understand information models, I have students read "What is an Information Model and Why do You Need One?" (Hackos 2002b). According to Hackos (2002b), an information model describes all of the content for your project: it is "an organizational framework that you use to categorize your information resources" (p. 3). Basically, the information model divides the content into logical chunks of information, labels these chunks with semantic tags, and organizes the chunks into structures. In the following example (Table 6.2), the student used *Excel* to create a model that shows all the information chunks a department needs to track and indicates where that information resides or will reside on the website.

The key to this assignment is a clear, tight focus. Often at this stage of the project, students discover that they do not have time to create an entire information model—or more likely series of information models—for their projects. If there is a major limitation to the course this is it: we do not have time to dedicate to information models. Ideally, I would teach a semester-long class on information modeling.

Content Management Wrap-up

Finally, I ask the students to read Batova and Andersen's (2017) "A Systematic Literature Review of Changes in Roles/Skills in Component Content Management Environments and Implications for Education." This article offers a comprehensive overview of the knowledge and skills technical communication students need to function in today's component content management (CCM) environments. These skills include project management and leadership, collaboration, content marketing, technology, and global workflow. Therefore, we need to provide curricula that focus on CCM methodologies, processes, and technologies (p. 190). As educators, we need to

1. *Build academy-industry education alliances*
   In their literature review, Batova and Andersen (2017) found multiple calls for technical communication curriculum that focuses on topic-based information development and single sourcing, but only a handful of authors describe their curriculum. They say:

   > To design or not to design curricula derived from workplace practice can no longer be a matter of theoretical perspective; entry-level technical communicators are now expected to have a baseline understanding of the methodologies, processes, and technologies that support CCM, and they are expected, at the very least, to be able to produce modular, reusable content.
   >
   > *(p. 191)*

TABLE 6.2 Information Model for a Website

Required							*Optional*	
General			*Contact*		*Forms*	*Contact*		*Graphics*
Shop Name	*Shop Responsibilities*	*Duties, Areas of Expertise*	*Phone #*	*Email*	*Situational*	*Personal Cell #*	*Personal Email*	*Situational*
Business Center								
Information Services								
Campus Planning & Construction								
Carpentry								
Custodial								
Davis & Auxiliary Campuses								
Electrical								
Electronic Systems & Repair								
Energy Management								
Fire Marshal								
Fleet Management								
FM Financial & Human Resources								
FM Systems & IT Support								
HVAC/R								
Key & Lock								
Landscaping								
Paint								
Plumbing								
Preventive Maintenance								
Real Property								
Sustainability								
Vehicle Repair								
Warehouse								
Website Location	Main Body	Main Body	Contact Bar	Contact Bar	Info Bar	Contact Bar	Contact Bar	Main Body

Note. Christopher Sessions, "WSU Facilities Management Website Information," submitted to ENGL 4110 on April 12, 2018.

2. *Rebrand the value of writing skills*
   Students need to understand CMS, XML, modular and reusable content, and so on. We must shift from document-based to topic-based writing.
3. *Integrate CCM in the undergraduate curricula*
   The authors cite several approaches: Robidoux's structured writing modules (2008); McShane's (2007, 2009) XML in CM; Stewart's (2014) XML in new media; and Sapienza's (2007) single sourcing.
4. *Decide on specialist or generalist technical communication education*
   There are no clear answers on this call to action. But clearly graduates who can model and write structured content will be strong candidates for content author and editor positions in CCM environments. Educators need to help students adapt, innovate, and learn.

Our wrap-up reading and response coincide with the completion of the students' CM projects. They develop the content using the tool(s) of choice, then submit that content along with a cover memo that reiterates the purpose of their CM project and outlines the content included (Step 4).

The CM projects vary widely. For instance, the Website project (noted previously) consisted of several deliverables: the information model in *Excel* plus a series of forms in *Google Forms* and *Word*. What I find most compelling about their final projects is not the end product necessarily but the description of their processes—how students analyzed the problem and discovered a solution. Ultimately, they discover that there is much more to "content management" than they initially thought.

Conclusion

Almost a decade after XML became a standard, I argued that technical writers were uniquely situated to write XML code because they could create their own tags using a metalanguage built with semantic tags. Twelve years later, I argue that in spite of its automation (through various CM systems and WYSIWYG processing tools), XML remains a relevant language to teach students the rhetorical strategies involved in writing code, in organizing content, and in structuring information architectures. Moreover, the act of hand-coding reinforces the technical skills necessary for graduates to function in today's workplace. I am encouraged that Batova (2018), in "Work Motivation in the Rhetoric of Component Content Management," also argues that (XML) CCM is rhetorically complex, requiring technical acuity and semantic skill. Since I began teaching the class, I have consistently received feedback from students that encourages me to continue teaching these relevant skills. Here are some answers to the question "What did you find of most value in this class?" from student evaluations, 2012–2018:

- "Learning XML."
- "Learning XML is a great advantage, especially in the workplace."
- "It was great to learn XML. This is a skill that I will definitely use in a future career. The final project was a good way to prepare to use content management systems and will be good to use in my portfolio."
- "I thought that the XML information was very helpful and the idea of chunking information also very helpful."
- "I love that I learned skills that employers are expecting from potential job applicants such as XML and HTML."
- "Prior to the class, I had a little experience with using HTML for content management. This class helped me become comfortable with HTML, CSS, and XML for content management. The class also inspired me to learn some JavaScript on my own. I think these are valued skills to have in a technical writing career."
- "I loved the coding."
- "The opportunity to work with information models and HTML."

Although I continue to update the readings, assignments, discussions, and tools—as the field of CM evolves—I remain committed to introducing students to the concepts behind CM and, in particular, hand-coding with XML.

References

Ament, K. (2002). *Single sourcing: Building modular documentation.* Norwich, NY: William Andrew.

Andersen, R. (2014). Rhetorical work in the age of content management: Implications for the field of technical communication. *Journal of Business and Technical Communication, 28*(2), 115–157. doi:10.1177/1050651913513904.

Applen, J.D., & McDaniel, R. (2009). *The rhetorical nature of XML: Constructing knowledge in networked environments.* New York, NY: Routledge.

Batova, T. (2018). Work motivation in the rhetoric of component content management. *Journal of Business and Technical Communication, 32*(3), 1–39. doi:10.1177/1050651918762030.

Batova, T., & Andersen, R. (2017). A systematic literature review of changes in roles/skills in component content management environments and implications for education. *Technical Communication Quarterly, 26*(2), 173–200. doi:10.1080/10572252.2017.1287958.

Batova, T., Andersen, R., Evia, C., Sharp, M.R., & Stewart, J. (2016, September 23–24). Incorporating component content management and content strategy into technical communication curricula. *Proceedings of SIGDOC Conference '16*, Silver Spring, CO.

Castro, E. (2001). *XML for the world wide web: Visual quickstart guide.* Berkeley, CA: Peachpit Press.

Clark, D. (2008). Content management and the separation of presentation from form. *Technical Communication Quarterly, 17*, 35–60. doi:10.1080/10572250701588624.

Cowan, C. (2008). *XML in technical communication*. Croydon: Institute of Scientific and Technical Communicators.

Ethier, K., & Houser, A. (2001). *XML weekend crash course*. New York, NY: Hungry Minds.

Glushko, R.J., & McGrath, T. (2005). *Document engineering: Analyzing and designing documents for business informatics and web services*. Cambridge, MA: MIT Press.

Hackos, J. (2002a). *Content management for dynamic web delivery*. New York, NY: John Wiley & Sons.

Hackos, J. (2002b). What is an information model and why do you need one? *Gilbane Report*, *10*(1, February), 1–32.

Hackos, J. (2008, April). When is minimalism the best course? *CIDM Information Management News*.

Hart-Davidson, W. (2010). Content management: Beyond single sourcing. In R. Spilka (Ed.), *Digital literacy for technical communication: 21st century theory and practice* (pp. 128–143). New York, NY: Routledge.

Hipps, J.B. (2016, May 21). To write better code, read Virginia Woolf. *New York Times*.

Lewis, M. (2012, February). Learning to manage your content. *CIDM Information Management News*.

McCarthy, J.E., Grabill, J.T., Hart-Davidson, W., & McLeod, M. (2011). Content management in the workplace: Community, context, and a new way to organize writing. *Journal of Business and Technical Communication*, *25*(4), 367–395. doi:10.1177/1050651911410943.

McShane, B.J.G. (2007, April). How to teach XML: A brief tutorial. *Intercom: The Magazine of the Society for Technical Communication*, 20–21, 29.

McShane, B.J.G. (2009). Why we should teach XML: An argument for technical acuity. In G. Pullman & B. Gu (Eds.), *Content management: Bridging the gap between theory and practice* (pp. 73–85). Amityville, NY: Baywood.

O'Keefe, S. (2009, September/October). The ABCs of XML. *Intercom*, 25–26.

Pringle, A. S., & O'Keefe, S. (2003). *Technical writing 101: A real-world guide to planning and writing technical documentation*. Research Triangle Park, NC: Scriptorium Press.

Pullman, G., & Gu, B. (2008). Guest editors' introduction: Rationalizing and rhetoricizing content management. *Technical Communication Quarterly*, *17*(1), 1–9. doi:10.1080/10572250701588558.

Pullman, G., & Gu, B. (2009). *Content management: Bridging the gap between theory and practice*. Amityville, NY: Baywood.

Robidoux, C. (2008). Rhetorically structured content: Developing a collaborative single sourcing curriculum. *Technical Communication Quarterly*, *17*(1), 110–135. doi:10.1080/10572250701595652.

Robles, V. D., & Frith, J. (2018, August 3–5). Developing a content strategy course and interdisciplinary skills: A teaching case. *Proceedings of SIGDOC Conference '18*, Milwaukee, WI.

Rockley, A. (2001). The impact of single sourcing and technology. *Technical Communication*, *48*(2), 189–193.

Rockley, A. (2002). *Managing enterprise content: A unified content strategy*. Berkeley, CA: New Riders Press.

Sapienza, F. (2007). A rhetorical approach to single-sourcing via intertextuality. *Technical Communication Quarterly*, *16*(1), 83–101. doi:10.1080/10572250709336578.

Stewart, J. (2014). Implementing an XML authoring project in a new media course. In 2014 IEEE *International Professional Communication Conference Proceedings* (pp. 1–7). Retrieved from http://ieeexplore.ieee.org/stamp/stamp.jsp?tp=&arnumber=7020386.

Thomas, S., & McShane, B. J. G. (2007). Skills and literacies for the 21st Century: Assessing an undergraduate professional and technical writing program. *Technical Communication, 54*(4), 412–423.

7

A RHETORICAL APPROACH TO TEACHING SOCIAL MEDIA TOOLS

Elise Verzosa Hurley

ILLINOIS STATE UNIVERSITY

Amy C. Kimme Hea

UNIVERSITY OF ARIZONA

Chapter Takeaways

- An understanding of social media tools (SMTs), their functions, and how they can be used in social media content management.
- An understanding of how reach as a heuristic can guide social media management and marketing.
- An understanding of how to integrate SMTs pedagogically in professional and technical communication courses.

As you know from your own experiences and work with students, social media are now a well-established aspect of our daily lives, and they are also a central component of the communication strategy of numerous organizations. In this chapter, we explore the ways in which SMTs—*Sprout Social*, *Buffer*, and *Hootsuite*, among others—can be used pedagogically in the professional and technical writing classroom. SMTs have been designed to connect several social media sites such as Facebook, Twitter, Instagram, to name a few, into a single platform to orchestrate social media content management. Although students might be frequent users of social media, they likely have less experience considering social media in the classroom or even in their future lives as professionals (Vie, 2008, 2017; Kimme Hea, 2011; Singleton & Melonçon, 2011; Verzosa Hurley & Kimme Hea, 2014). Here, we outline three professional and technical writing assignments to introduce and deepen student understanding of SMTs and provide a conceptual framework of reach (Pearson, 2011) as a means to contextualize these assignments.

Because SMTs are relatively new tools and are not traditionally conceived of as content management systems (CMSs), we first explain how they function and describe their affordances. Then, we briefly situate the content management capabilities of SMTs in relationship to more traditional CMS approaches, arguing that SMTs introduce an additional role beyond the normal content management for a social media content manager. Finally, we introduce three assignments with the aims of developing rhetorical deployments of SMTs as part of a professional and technical writing classroom. When we teachers of professional and technical writing ask students to think of SMTs as integrated with an organization's communication strategy and its brand, students can use this rhetorical approach to think critically about content development and management, which are key to effective social media deployment.

SMT Functions and Affordances

As previously mentioned, SMTs such as *Sprout Social*, *Buffer*, and *Hootsuite* are integrated social media management tools that allow users to create, schedule, publish, monitor, and curate content across multiple social media platforms to drive user engagement for content marketing or brand identification purposes. Although all three platforms allow individual users access to free plans to manage personal social media accounts, *Sprout Social*, *Buffer*, and *Hootsuite* are specifically targeted to professional users—social media managers—who want to grow the reach of their organization's influence by tapping into specific affordances of individual social media platforms to build cohesive brand identification and affinity across varied (though connected) social media networks. Key to building social media presence, brand identification, and engagement, then, is maximizing the reach of social media messages and SMTs provide the necessary tools for social media content managers to do so.

Product developer Bob Pearson (2011) argues that in order for organizations to succeed in the age of social media, they must tap into the affordances of Web 2.0 by anticipating users' social and browsing habits, preemptively reaching audiences through dynamic content and content marketing, and facilitating brand identification and affinity through user engagement (p. 5). Elsewhere, we have written that reach is a useful metaphor when teaching technical communication students the affordances of composing with social media (Verzosa Hurley & Kimme Hea, 2014), and we continue to believe that a rhetorical understanding of reach can help students learn to navigate the affordances of SMTs. Further, we believe that teaching students to become familiar with creating and curating content through these tools is vital because, as Frith (2014) posited, "technical communicators are uniquely suited to step in to a variety of social media management positions" (p. 182). In the following sections, we discuss two primary features of SMTs—dashboard and analytics—which allow social media content managers to create, curate, respond to, and drive social media content. In doing so, we focus less on specific how-to guides for each individual SMT because platform technologies and their attendant interfaces can change quickly; rather, we believe that it is more useful to provide an overview of SMTs' capabilities so that you as individual instructors can become familiar with their functionality and, perhaps, choose one to integrate into your classroom pedagogies and design your own projects alongside those we are offering here.

Dashboards

Although SMTs have varying dashboard interfaces, they all provide users with tools to (1) connect social media networks (i.e., Twitter, Facebook, Pinterest, Instagram, LinkedIn) to its main platform; (2) create multimedia content posts that are customized for each social network; (3) schedule the frequency of content posts; (4) plan and schedule discrete content variables ahead of time as part of managing a larger social media marketing plan or campaign; (5) respond to user feedback on each individual social network; and (6) access user analytics to learn *how* and *when* to maximize content reach and user engagement.

Similar to previous discussions on single sourcing whereby authors are tasked to create content that can then be distributed in different products to diverse audiences (Eble, 2003), SMTs allow social media managers to create content that originates from the specific SMT's platform so that it can be simultaneously distributed to multiple social media networks. While traditional single sourcing emphasizes the need for the individual writer to write and structure content for understandability in different contexts and then accurately label and categorize content for reuse (Butland, 2001; Eble, 2003), SMTs have the capability to optimize content for each individual social media network. For example, *Buffer*'s platform includes a "Tailored Posts Composer" that allows social media managers to create the same content specifically optimized for different social media networks such as Facebook, Instagram, and Twitter. The social media manager would simply create content in *Buffer*—complete with a written message and some combination of images, videos, or GIFs, etc.—and *Buffer*'s Tailored Posts Composer will suggest condensing or expanding the written message based on the character limits of each social media network, suggest a high-definition image for Instagram or a more detailed video for Facebook, and recommend a pithy GIF for Twitter (all of which take advantage of the affordances of each social media network—e.g., the visual appeal of Instagram, the direct community engagement of Facebook, the brevity of Twitter). In addition to integrating mentions across all social media networks and suggesting the appropriate amount of trending hashtags (for example, Instagram posts tend to have more hashtags per post versus Twitter's one or two hashtags per tweet), SMTs also provide prompts to consider social media composition.

As such, social media managers need a working knowledge of the potentials and constraints of individual social media networks in relation to their organization's overall goals and content marketing strategy. In addition to the primary content users might create, for example, *Hootsuite*'s interface allows social media managers to create streams that uncover new and related content that they might wish to repost, to organize content by tags and usage statistics in a content library, and to share these resources with anyone on their social media management team. Such features not only emphasize the content an individual user creates but also the wealth of continuous content constantly created across social media networks that may then be reused and repurposed to fit the individual organization's goals and strategy. The concept of reaching certain audiences on the range of social media platforms is part of the dashboard in the SMT.

Social media use often depends on temporality or the timing of delivering content messages to the right social media network and the target audiences at the right time. Thus, SMTs are also project management tools as much as they are content management tools. Because organizations often depend on an overall campaign or strategy that grounds their ethos in specific strategies that facilitate customer identification and affinity, SMTs allow social media managers to create, plan, and schedule release of content to meet business and brand goals. For example, *Sprout Social*'s interface allows social media managers to map content themes onto a calendar, giving them a holistic view of a marketing campaign or strategy. Thus, scheduling and planning upcoming content is also a key feature of many SMTs. Social media managers, then, must be able to plan a campaign or strategy in advance and create or curate scheduled content that drives the overall campaign.

Analytics

In addition to the dashboard capabilities that allow social media managers to create, plan, curate, and distribute content, SMTs also come with built-in analytic features. These features are often connected to the brand identification of the organization implementing the platform. Rather than consider SMTs as merely tracking a company or individual, SMT analytic features are most used to create relationships with dispersed, connected users of different social media sites. These connections are understood as central to the development of relationships with current or potential customers or users. In marketplace models, the relationship is cultivated to create loyal customers, those who will not only buy the product once but will become loyal to it in the future. It is not merely the specific product circulated, but the brand, or the whole set of relations evoked by the organization or product, including feelings and perceptions that become attached to the overall brand. Branding is something of concern to companies and to selling products and services, but non-profits, cities and towns, and even individuals also might invest in establishing a brand. In Brown's (2016) extended introduction to the complexity of defining brands and branding, he suggests that at its core, "[b]randing differentiates. Branding separates" (p. 22). For Brown, despite the contested definition of *branding*, the clear purpose of building a brand is to promote a particular product, service, or even place as the best choice for a consumer.

Consider any celebrity who might engage social media to interact with fans. These engagements—this reach to the public—are not just in one direction with most social media, but the fans also can engage virtually with their favorite actor, for example. Thus, SMTs become a means to synthesize the interactions and to build that affinity (and identification), and cultivate a brand. A recent example of how a company used social media to influence its brand was iHop. The International House of Pancakes, a chain restaurant of Dine Brands, is known for its breakfast menu and, unsurprisingly, its pancakes. The company, however, deployed an interesting strategy to reframe its brand by announcing a name change to iHob, or International House of Burgers. It circulated this name change through several social media networks, and in Los Angeles, at its Hollywood store, it changed its store sign to iHob. NPR, the *Washington Post*, and other news outlets interviewed the CEO about the change and reported on the move as serious business news. Soon, McDonald's, Burger King, Wendy's, and other restaurant chains were sending tweets about the change, making fun of the company for its lack of savvy to move away from its brand and an association with breakfast, with dinner culture, and with unpretentious food. Wendy's tweeted, "[n]ot really afraid of the burgers from a place that decided pancakes were too hard." What came from all the clamour and social media chatter, however, and from companies and individuals, was exactly the buzz that this stunt garnered: attention to its burgers, its lunch menu, its ability to use humor, and its practical joke on other companies, which all meant allowing social media to light up with jokes, but indeed, the message circulated broadly across social media sites. The branding stunt was seen as a marketing coup, with marketing gurus applauding iHop for breaking through all the other messaging to eventually "double its comparable burger sales since before the promotion, according to its parent Dine Brands Inc." (Garcia, 2018, para 1).

We are uncertain if iHop used a specific SMT, but it can be assured that such a platform was tracking what was happening through mentions, likes, hashtags, photos, and other information, gathering insights on who was composing, on which social media sites, on where the buzz was growing, at which times, and on how the messages were sent via iPhone, laptop, and even from which IP addresses and locations. All these data are aggregated into visualizations that allow users of the SMT to determine their reach and to identify consumer segments and social media influencers (those individuals who have the most engagement with others).

One important consideration is that the analytics in SMTs are a differentiator across the most popular platforms. *Sprout Social*, for example, provides a host of visualizations of already digested data for its social media managers to act upon. These analytics display the who, what, where, when, and even how of the social media users' engagements with the sites that have been integrated into the SMT. This set of data tracks the ways in which individual users engage with the brand, including such common engagements as rating of posts (likes or stars), recirculation of a post (retweets or shares), tagging (hashtags and photo tags), or other means to demonstrate action in a social media venue. In this view, the actions across the social media sites are represented to the social media manager to develop or recirculate content or to feedback into the communication strategy.

SMTs and New Roles for Content Managers

Although SMTs are not content management tools in the traditional sense, they do enable social media content managers to coordinate the use of various social media accounts through a single tool that allows for dissemination of content, analysis of engagement, and coordination of messages across social media platforms. And although a social media manager might be authoring content to distribute across social media through SMTs, her role is much more one of orchestration, or laying out the brand and communication strategies necessary to make all the different social media sites play in concert with one another. Once a social media manager subscribes to an SMT and connects individual social media accounts to the SMT platform, she can see all the different tags, mentions, and likes in relation to trending keywords and concepts that make an organization's social media presence or "brand" distinctive. Keeping with the music theme, imagine that a social media manager is working as an SMT consultant for a new classical pianist. As a task, the social media manager would connect the pianist's social media sites—Facebook, Instagram, Pinterest, and even a YouTube channel—and access them through an SMT like *Buffer*, *Hootsuite*, or *Sprout Social.* Using the SMT dashboard allows the social media manager to trace the circulation of the pianist's social media content, including mentions, tags, and repostings, across all the individual social media accounts. The social media manager can also see that the pianist has a fanbase in a certain region, of a certain age, and even that prefers a certain social media platform or type of social media content, by accessing SMT analytics. From this information, the manager can begin to create new social media messages that engage these fans, but she also can use the fan messages about this artist—and even other classical pianists—to broaden the reach of the artist to grow her community of followers and fans. This example gives you a sense of the ways that an SMT provides a one-stop view across social media and allows the social media manager to orchestrate the engagement of social media users and their own user-generated content. Thus, teaching students how to use SMTs in professional and technical writing contexts allows them to not only understand how to create content for social media but also to develop strategies for broadening their content's reach and circulation.

Using SMT in the Classroom

Now that we have explained the capabilities of SMTs and why they should be viewed as content management tools, we turn to how SMTs can be used in an introductory professional and technical communication course that asks students to analyze and compose social media content within specific contexts of use. The assignments we include here build from analysis to production, prompting students to deepen their understanding of creating content for social media. Early in the semester, instructors should ask students to describe their experiences with social media either through a brief informal questionnaire or in-class writing or discussion prior to introducing concepts such as reach and engagement within and across individual social media platforms.

Social Media Analysis

This first assignment serves as an introduction to understanding and analyzing how organizations create, share, and manage content across different social media platforms. This assignment allows students to critically examine an organization of their choice and how they use social media to construct brand identification and engagement with specific audiences. The project description says:

> This assignment asks you to critically examine the social media presence of an organization, business, or company of your choice by analyzing its social media presence through at least two specific social media platforms (i.e., Facebook, Twitter, Instagram, Snapchat, YouTube) at the same time you learn a key technical and professional communication genre: the informal report. This project includes, then, an informal report and class presentation.
>
> Essentially, this assignment asks you to research the following questions:
>
> - Who is the organization, business, or company? What is their mission and who are the audiences to which they cater?
> - What social media platforms do they use? Which platforms have you chosen to focus on, and why?
> - How many followers/fans does the organization, business, or company have? Who are the followers/fans? How do you know?

- What types of social media activity does the organization, business, or company post (i.e., text, photos, videos, a combination of all three)? In what ways are the types of social media activity similar or different across platforms? What social media activity strategies do they use (frequency of posts, reposts, shares, tags, hashtags, etc.)?
- What kinds of interactions do they have with their followers/fans? How often do they interact with them? In what ways are the interactions similar or different across platforms?
- What kind of ethos or image does the social media presence of the organization, business, or company present to its audience? Comment on its effectiveness.

Prior to assigning the project, instructors should walk students through several in-class discussions that focus on examples of the social media presence of various organizations (the social media presence of your institution is a great place to start) and their strategies for engaging audiences. It is important for instructors to use examples from different kinds of organizations/people (e.g., a government organization, a business or company, a non-profit institution, a celebrity) so that students can develop a sense of the different purposes for having a social media presence as well as the various audiences they target.

After modeling several examples in class, students can begin to conduct individual research on the organization of their choice to write the informal report. In addition to the informal report, students should also present their findings to the class. The variation in organizations and social media strategies that students choose for this assignment serves as the basis for an excellent class discussion on effective social media communication across organizational contexts.

Social Media Technology Quick Reference Card

This SMT Quick Reference Card (QRC) assignment develops students' facility with SMTs, but it also allows them to work across the class to build rapport and to collaborate. That said, the SMT QRC project could be developed either individually or collaboratively, but we will discuss it here as a team-based assignment. Keep in mind that this assignment will discuss the QRC deliverable as hard-copy product that could be distributed physically to other class members, but it could also be easily stored in a shared space for easy student access. This assignment lays the groundwork for fellow students to understand SMT functions, but it also could be easily adapted to share with other audiences, which would simply require different discussions of design, purpose, and audience. The project description says:

As the previous assignment provided critical awareness of the ways social media are part of our daily lives, this assignment situates the deployment of SMTs in relationship to the class members with the aims to build student collective technology learning and awareness and to allow students to employ SMTs for a client project or other course project in the class. This project includes three deliverables: documentation memo, design template, and SMT QRC, scoped appropriately to the class needs. It is important that as the course instructor, you situate this project in relationship to future SMT use in the class. For example, will students be crafting a social media campaign? Will they be teaching a client to use an SMT dashboard? This project relies on the course to be the context and help define the audiences and purposes for the QRC.

The students will work together to assess student comfort and knowledge of SMTs and other aspects of social media. Some of the questions students may consider as they strive to understand their fellow users include:

- Among our class members, what is the range of technical expertise?
- What is their general comfort level with learning new technology?
- How many of your peers are familiar with the specific SMT or social media you are documenting?
- What is your audience's "usual" approach to learning unfamiliar technologies?
- What particular aspects of the SMT or social media do your peers want to learn?
- What possible course-related tasks will they need to complete with the SMT or social media?
- How will your quick reference card address the audience learning goals and task-related needs?
- Based upon the audience expertise, comfort, goals, and needs, what specific aspects of the program will you document in your card?
- What is your own background and experience with the SMT or social media you will be documenting? What steps will you need to take to familiarize yourself with the technology?

These questions can be adapted to fit the particular class goals or related assignments.

After completing a documentation memo that assesses student class members' audience needs, students complete, in their groups, a design memo. This memo is a means to map out the QRC design, including visual design aspects, the user needs, usability and accessibility, and the scope of their QRC. One of the challenges, as you might anticipate, will be the scope of how much an SMT QRC can address. These SMT QRC cards will be shared across the teams as a vehicle to learn more about the specific technologies. In this way, the students are gaining facility with the technologies to produce the QRC as they also will be using the cards to assist with other projects in the course.

Finally, this project requires students to complete the SMT QRC as its final deliverable. These QRCs will be useful to other members of the class, and as they are working in teams, they also are helping one another overcome limits of their own knowledge to produce the project. This working together also assists them with future collaborative projects. You may elect to have teams provide you with one full-color printed and lamented version, which you can ask to use in future classes as an example of the project or part of a knowledge bank you provide to other classes learning to use SMTs.

Social Media Campaign

This team-based capstone project works especially well as a client project where students are asked to develop and launch a social media campaign that addresses the needs of a specific organization. The best clients for this project are those that have existing social media accounts that may not be regularly managed or lacking in engagement (think student or departmental organizations that do not have a designated social media content manager). By building a relationship with a partnering client and adopting a consultant stance, students are able to put their social media content management skills to use for real audiences. The project description says:

> This assignment asks you and your team members to establish a client–consultant relationship with a campus organization to research, propose, and launch a social media campaign that addresses the needs of your partnering client. You will research your client's organization to determine their social media needs, write a proposal that outlines a specific social media campaign and, with your client's approval, launch the campaign using *Buffer*. Two weeks after your team's campaign is live, you will synthesize the data from *Buffer*'s analytics and write a memo about the campaign's effectiveness and areas for improvement. The deliverables for this project include a proposal, social media campaign launch, and memo.
>
> Instructors may choose specific client(s) for students to work with or students may choose their own clients. In either instance, instructors should guide students through contextual research on their partnering client's organization to determine the client organization's social media needs. A face-to-face interview with the client and web research on the social media presence of similar organizations allow students to gain understanding of the communication needs of their clients; thus, we recommend that instructors walk through writing effective interview questions and

conducting interviews with their students. Once students have completed the necessary research, they can begin crafting a proposal (addressed to the client) that outlines a social media campaign that will meet their partnering client's communication needs. Scaffolding is crucial for this project and we find that the following heuristics centered on maximizing reach can complement the research portion of the project and aid students in writing their social media campaign proposals:

- Who is the target audience you are trying to reach?
- What are their interests, values, and characteristics?
- What are the interpersonal and social relationships among the target audiences?
- Which kinds of social media platform are effective in building social relationships among the target audiences and with the brand identity?
- What kinds of content and modes of delivery will likely be effective for both the target audiences? How is this content constrained or enabled by particular social media platforms?
- What visual and digital media might be used to enhance engagement through specific social media platforms?
- What hashtags might be composed in order to tag content for particular audiences?
- When do the posts need to be made live in order to reach target audiences?
- How might likes and mentions of previous customers be reused to create and tag new content?

Once the students have gained the approval of their partnering client, they can then launch the social media campaign using the SMT you, as the instructor, want them to use. The types of social media campaigns will vary depending on the specific client context, though students may likely advertise a specific event or cause. In so doing, they will also create a posting schedule, produce images and/or other multimedia, create hashtags, and design posts meant to target and engage with specific audiences. Students should monitor their social media campaign and, after two weeks, will synthesize the results of their campaign using SMT analytics, noting the campaign's strengths, constraints, and potential for improvement outlined in a detailed memo. The audience for this memo is both you and the partnering client, who can use the information to increase social media engagement or for any future social media campaigns.

As Vie (2017) noted in her extensive study of technical communication praxis and the integration of social media, the desire, or need, to teach social media is integral to the field of technical communication and the reality of technical communication students who need the skills of distributed work to thrive in contemporary workplaces (Spinuzzi, 2007, 2014; Faris & Selber, 2013; Ferro & Zachry, 2014; Pigg, 2014). In closing, we believe that SMTs as content management tools can also be vehicles to engage in broader discussions about the nature of technical communication in the social media landscape, analysis of privacy and laws related to social media in the United States and abroad, the role of influencers, the concept of brands and branding, and the ethical and rhetorical dimensions of SMTs and social media in general. As technical communication spans these new technological ecologies, we posit that instructors should consider integrating not only the specific uses of SMTs but also the broader questions about rhetorical deployment of these tools in their classroom pedagogies.

References

Brown, S. (2016). *Brands and branding*. Thousand Oaks, CA: Sage.

Butland, P. (2001). Introduction to single source. *Intercom*, *48*(2), 23–27.

Eble, M. (2003). Content vs. product: The effects of single sourcing on the teaching of technical communication. *Technical Communication Quarterly*, *50*(3), 344–349.

Faris, M.J., & Selber, S.A. (2013). iPads in the technical communication classroom: An empirical study of technology integration and use. *Journal of Business and Technical Communication*, *27*(4), 359–408. doi:10.1177/1050651913490942.

Ferro, T., & Zachry, M. (2014). Technical communication unbound: Knowledge work, social media, and emergent communicative practices. *Technical Communication Quarterly*, *23*(1), 6–21. doi:10.1080/10572252.2014.85084.

Frith, J. (2014). Forum moderation as technical communication: The social web and employment opportunities for technical communicators. *Technical Communication*, *61*(3), 173–184.

Garcia, T. (2018, November). IHOP's name-change stunt helped double its burger sales. *MarketWatch*. Retrieved from www.marketwatch.com/story/ihops-name-change-stunt-helped-double-its-burger-sales-2018-11-03.

Kimme Hea, A.C. (2011). Rearticulating Web 2.0 technologies: Student constructions of social media in community projects. In M. Bowdon & R. Carpenter (Eds.), *Higher education, emerging technologies, and community partnerships: Concepts, models, and applications* (pp. 235–244). Hershey, PA: IGI Global.

Pearson, B. (2011). *Pre-commerce: How companies and customers are transforming business together*. San Francisco, CA: Jossey-Bass.

Pigg, S. (2014). Coordinating constant invention: Social media's role in distributed work. *Technical Communication Quarterly*, *23*(2), 69–87. doi:10.1080/10572252.2013.796545.

Singleton, M., & Melonçon, L. (2011, June). A social media primer for technical communicators. *Intercom*, *58*, 2–9.

Spinuzzi, C. (2007). Guest editor's introduction: Technical communication in the age of distributed work. *Technical Communication Quarterly*, *16*(3), 265–277. doi:10.1080/10572250701290998.

Spinuzzi, C. (2014). How nonemployer firms stage-manage ad-hoc collaboration: An activity theory analysis. *Technical Communication Quarterly*, *23*. doi:10.1080/10572252.2013.797334

Verzosa Hurley, E., & Kimme Hea, A.C. (2014). The rhetoric of reach: Preparing students for technical communication in the age of social media. *Technical Communication Quarterly, 23*(1), 55–68. doi:10.1080/10572252.2014.850854.

Vie, S. (2008). Digital divide 2.0: "Generation M" and online social networking sites in the composition classroom. *Computers and Composition, 25*, 9–23.

Vie, S. (2017). Training online technical communication educators to teach with social media: Best practices and professional recommendations. *Technical Communication Quarterly, 26*(3), 344–359. doi:10.1080/10572252.2017.1339487.

PART III

Tasks

8

INCLUSIVE AUDIENCE ANALYSIS AND CREATING MANAGEABLE CONTENT

Carleigh Davis

MISSOURI SCIENCE & TECHNOLOGY

Michelle F. Eble

EAST CAROLINA UNIVERSITY

Chapter Takeaways

- Describes a heuristic that instructors can use to teach inclusive audience analysis for creating reusable content.
- Integrates a case-based approach to technical communication instruction that introduces specific theoretical considerations for rhetorically effective writing for content management systems (CMSs).
- Provides sample scenarios that can be used to help students practice writing for a variety of audiences using a CMS.

This chapter sets out to answer the following pedagogical question: how can instructors help students account for a range of multiple audiences, stakeholders, contexts, and purposes when teaching students to write content? In other words, how do we teach students to conduct audience analysis for writing and reusing content in a content management system? This chapter advocates for the practice of inclusive audience analysis as a means of teaching effective and ethical practices for creating content for use in a CMS. To help instructors encourage inclusive audience analysis with their students, we introduce a process and method, which we call a heuristic, that instructors can use to help students think about, write, and deliver audience-focused CMS writing that can be applied in a variety of scenarios and enables students to approach this learning systematically. The heuristic is comprised of six steps that can be integrated into a case-based approach to technical communication instruction when the instructor wants to give students practice with rhetorically effective writing for a CMS. The heuristic consists of the following steps:

1. Analyze the active interfaces maintained by the CMS and complete a content audit.
2. Understand the affordances, constraints, and conventions of the CMS.
3. Determine the needs and motivations of audiences and stakeholders.
4. Determine how the active interfaces work to meet the needs and motivations of audiences and stakeholders.
5. Create content based on audience and stakeholder needs and values within this system.
6. Use content creation to compensate for shortcomings in the system.

This chapter also offers a case-based example that illustrates the use of this heuristic in practice, which instructors can use as a generative example to help design their own formative projects and assessments for each step in the heuristic. Finally, the chapter concludes with two sample scenarios that can be used as cases to help students practice inclusive and rhetorically effective writing for a variety of audiences using a CMS.

As instructors and technical communication professionals, we are committed to making the interactions between content, technologies (including CMSs), and stakeholders apparent. We believe that when these connections are ignored or de-emphasized, it creates further opportunities for technical communication to cause unintended harm to communities as well as unanticipated breakdowns in communication. As such, this chapter emphasizes the rhetorical impact of CMS technologies on audiences and stakeholders and advocates that this relationship be taken into account when instructors teach effective writing for use in a CMS.

Writing, Audience, and Content Management

Over the last 15 years, CMSs have dramatically changed and complicated the ways we have to think about our pedagogical approaches to teaching audience analysis and understanding in technical communication. This is especially true when content may need to be used in a set of contexts for various audiences, stakeholders, and users. Technical and professional communication is always audience- or user-focused; this is something we emphasize to students as a characteristic that distinguishes technical communication from other kinds of writing (see *Technical Communication Today* by Richard Johnson-Sheehan, the current recommended text for the Society for Technical Communication certification, among others). This priority has led to an increase in scholarship on specific practices related to content management, including concepts and theories, that focus on human-centered design (Rose et al., 2018), user experience (Sun, 2012; Gonzales, 2018), and content strategy (Batova & Andersen, 2016; Clark, 2016; Gonzales, Potts, Hart-Davidson, & McLeod, 2016).

Traditionally, we teach audience analysis as human-centric; it is a process that involves determining the wants, needs, and motivations of human readers and users engaging with the content that a technical writer produces. This analysis then feeds recursively into the creation of content, all aspects of which should work towards meeting the needs of the intended audience. Students learn to research, create content, modify tone, and, of course, to design documents with these audiences and users in mind. Such rhetorically-aware teaching practices are laudable and help students account for a range of audiences, stakeholders, contexts, and purposes when they write content. CMSs require additions to these practices by demanding that technical communicators consider a series of content types/genres and delivery methods, rather than a single scenario in which readers will interact with a document. Some of the content may be separated from the original document for which it was written; it may also be expected to transfer effectively across a variety of audiences and stakeholders rather than applying consistently to a single set of primary, secondary, and tertiary audiences. In the spirit of social justice, it is important that technical communication instructors provide students with the rhetorical tools that can be used in these situations to account for inclusivity when many audiences and stakeholders are at play.

Drawing on Hart-Davidson, Bernhardt, McLeod, Rife, and Grabill (2007) we understand

> the practical work of content management (CM) [as] a form of reasoning, *phronesis*, that permits us to explore CM as a means to guide decision making about the creation of knowledge, the arrangement of information, the selection of tools, and the design of work practices associated with the making of texts.
>
> *(p. 10)*

In other words, we understand content management as both a contextual and rhetorical practice. As a result of viewing CMSs as rhetorical, we have to account for audience and users and their relationship to the content they need. Likewise, Clark (2008) defines a CMS in the following terms:

> a system that approaches the problem of content management by using markup, metadata, and tools to break documents into component parts, to a level of granularity (e.g., paragraph-level, sentence-level, word-level) set by organizationally defined information models, and labeling each part with metadata that describe its meaning and relationships to other content. The same content can then be automatically assembled in different genres, with different presentations, and in different media.
>
> *(p. 39)*

This definition resists the idea of a CMS as a specific digital program, instead affording CMSs a range of applications across a particular rhetorical environment. We would extend this definition to include the stipulation that content should always be written to meet specific audience or user needs. In this way, a CMS is first and foremost a technology used to solve the problem of communication across both audiences and contexts.

These definitions help us to understand the interconnected work of writers, who often appear isolated in CMS contexts, as reaching through and beyond content management tools to meet the needs of the contexts and audiences for which they write. Batova and Andersen (2017) argue that those teaching technical communication classes should expand our view of the rhetorical situation when working with a CMS because a CMS "requires technical communicators to examine audience, purpose, and context from a very different perspective" (p. 192). They likewise suggest "that students receive adequate instruction in writing structured content and analyzing rhetorical situations that move well beyond the writer–audience–subject relationship" (p. 192). This emphasis on an expanded view of the rhetorical situation places analytical skills at the center of writing usable and effective content. One way to consider inclusive audience analysis in these contexts is through the articulation of content that users need, because this articulation reframes the discussion about audience in relationship to the content and prioritizes the audience over the information or content.

As technical communication instructors, we are committed to social justice theoretical frameworks in our approach to technical communication research and pedagogy. An emphasis on audiences, users, and stakeholders is essential to this approach. To this end, we believe that new instructors of technical communication need to be increasingly vigilant in teaching students to be cognizant of the audiences and stakeholders who might be affected by technical communication and technical documents. Although it might be easier and more efficient to view both technical communication and the technologies to which it is tied as neutral and objective, we must recognize "that technologies and sciences are culturally-rich and thus informed by ideological agendas and uses" (Haas & Eble, 2018, p. 5). The recognition and analysis of the ideological agendas and uses of various technologies is at the core of our technical communication pedagogy because we want students to have the knowledge and skills to create content that is inclusive and ethical across contexts. We understand that what instructors choose to teach and how instructors choose to teach are "always already influenced by theories about teaching, learning, and communicating about science and technology. Thus, all teaching is ideological and political, even if we pretend it is not" (Haas & Eble, 2018, p. 7). Knowing this, we believe it is the responsibility of technical communication instructors to provide spaces for critique of, and reflection on, writing technologies. Such spaces help to make the implicit power structures of writing technologies and technical communication scenarios apparent, challenging the all-too-common obfuscation of the real-world effects that technical communication has on stakeholders.

This practice is especially important in writing scenarios where stakeholders are physically and ideologically distanced from the content produced by technical writers, as is the case when content is managed and distributed by a CMS. Furthermore, the use of a CMS naturally influences the means of distribution, and therefore the rhetorical impact, of the content to which it pertains. Teaching that encourages students to be inclusive in writing manageable content must pair theoretical approaches that foreground the social and ethical implications of content management technologies with opportunities for students to create content that meets the needs of their users within specific contexts.

Providing students with the analytical tools to build awareness and knowledge about creating manageable content is especially important when writing content with and for CMSs because technology and content are so inherently dependent on one another. This interdependence provides further opportunity to ignore the needs of stakeholders in favor of the needs of the CMS. To conduct effective audience analysis when writing with and/or for a CMS, students must consider the needs of both the end users and the CMS itself as relevant audiences. In this way, the audience for the content they create is both human and machine. Students must learn to explore the system itself, as well as users' expectations of that system and their motivations for engaging with it to create manageable content within its parameters.

CMS Ecologies Heuristic

In teaching students both content management and writing manageable content, we use the term "ecologies" to refer to the network of actors and interactions influenced by a CMS, including but not limited to the CMS itself. These ecologies include writers and readers/users, motivations, and technologies, as well as information. While somewhat metaphorical, we find this term useful as a way of emphasizing the balance of agency and naturally dynamic content at play in these networks. The term "ecologies" also draws on Memetic Rhetorical Theory (Davis, 2018), which we consider a productive tool for thinking about how information and content adapts to new environments. According to this theory, information must adapt to an ecology by fitting in with the various memeplexes (that is, groups of ideologies, technologies, actors, etc.) that make up that ecology.

We advocate the use of a heuristic or series of questions to help students think about, write, and deliver usable content in CMS ecologies to their users and audiences. In this heuristic, we imagine students as content developers writing in scenarios in which a CMS has already been established and is working well to meet the needs of the authors, audiences, and stakeholders. It is modeled after Ridolfo and DeVoss's (2009) concept of *rhetorical velocity*, which they define as a "strategic approach to composing for rhetorical delivery," that "refers to the understanding and rapidity at which information is crafted, delivered, distributed, recomposed, redelivered, redistributed" (para. 1). This heuristic asks students to (1) consider how to be inclusive of their primary audiences and various stakeholders for the content they develop, and (2) account for potential contexts where their content needs to be used or accessed. Ridolfo and DeVoss (2009) define *rhetorical velocity* in the context of inventive thinking for composing as "the strategic theorizing for how a text might be recomposed (and why it might be recomposed) by third parties, and how this recomposing may be useful or not to the short- or long-term goals of the rhetorician" (para. 1). Although the concept of reuse—and in this quote, "recomposing"—is common in CMS literature, we resist this term due to the inherent privilege it bestows on the first chronological iteration of a particular piece of content. Instead, in our heuristic, we encourage students to think about strategically composing for multiple audiences and iterations of the content (which may occur either synchronously or asynchronously) from the onset in order to be more inclusive. Our heuristic incorporates rhetorical velocity in that it asks students to be strategic and cognizant of the ways the content will be managed and distributed from the invention stage of the process onward.

Like the memetic rhetorical theory it is based on, this heuristic is memetic in nature, meaning that it relies on the adaptability and suitability of content to the technological and rhetorical environments in which it will be used. Together, these technological and rhetorical environments form ecologies to which new content may successfully adapt or ultimately fail. Audience analysis and contextual analysis feed recursively into one another in such a system, where students must think of content production as intrinsically linked with content delivery as a rhetorical act. Because content divorced from audience/delivery loses its significance, students who think of content objects as decontextualized entities run the risk of unethical treatment of certain users or excluding important users to create catch-all content. In this addition, a modular approach to content management can disregard users and thus an important part of accounting for the rhetorical situation.

The following heuristic consists of a series of steps an instructor can use to help guide students through a process that asks them to think about how content will successfully adapt in specific technological ecologies.

Step 1: Analyze the Active Interfaces Maintained by the CMS and Complete a Content Audit

Creating content for an existing CMS means stepping into an environment that has established boundaries and expectations. The content that students create must be compatible with these rules and expectations to be usable for their intended audiences and users. Therefore, to begin the process of creating content for use in a CMS, students need to gain a thorough understanding of the uses and capabilities of the CMS they are working with. Instructors should frame discussions of the existing environment of a CMS as a way of understanding the connections among form, function, content, and audience by characterizing the CMS environment as an ongoing conversation between writers and their audiences facilitated by mediational and technological tools.

The first step in developing this understanding is for students to become familiar with the content already distributed through the CMS—its characteristics, conventions, topics, and purposes. We refer to the means and media of distribution within the CMS as the *interface*, which, in this context, we define as the way in which the audience/end users receive the CMS.

Step 2: Understand the Affordances, Constraints, and Conventions of the CMS

The next step is for students to analyze how the affordances of the CMS facilitate communication using these characteristics. After all, content and form are intrinsically linked in rhetorical practice; it is not enough for students to understand what is communicated; they also must understand how that communication occurs, both from the perception of audiences who engage with the interface and from the perception of writers who input the content into the CMS.

Together, Steps 1 and 2 of this heuristic encourage students to become familiar with the rhetorical environment to which their content will need to adapt.

Step 3: Determine the Needs and Motivations of Audiences and Stakeholders

Like all technical communication, the creation of content for use in a CMS should always be focused on the needs of a particular target audience, or target audiences, as well as potential stakeholders. Therefore, students must spend time working to understand the needs and motivations of the eventual readers and users of the content they create. These needs are likely specific to the relationship between the readers and the institution publishing the material, so students should start by identifying that relationship. Next, students should break users and stakeholders into specific groups based on their needs and motivations when engaging with published content. This will likely require significant research, both within the organization and using demographic resources relevant to the kind of content the CMS manages. These specific groups might be based on user roles or might be based on other characteristics, but the focus should be on being as inclusive as possible.

Step 4: Determine How the Active Interfaces Work to Meet the Needs and Motivations of Audiences and Stakeholders

To reinforce the rhetorical connection between content and delivery, students must understand why the interface features that they have identified have been successful in meeting the needs of the audiences and stakeholders for the CMS (this is, of course, assuming that the CMS is well designed and has been successful in this respect). Students need to cultivate an understanding of why various interfaces are used for this purpose, by this organization, etc., based on the needs of the intended audiences. This understanding may be evident, or may require further research. In short, they must answer the question: what is it that makes the genres, social media sites, web platforms where the information might appear work for this audience? The answers to this question should yield a series of aligned interface features and audience needs. These combinations of interface features and audience needs are the memeplexes that define the CMS ecology.

Step 5: Create Content Based on Audience and Stakeholder Needs and Values within This System

As in all technical communication contexts, content creation in CMS ecologies should meet a specific need relating to communication between the writer and the audience or user. This need may be consistent with past communications or may be related to new events or goals. Regardless, new content must be consistent enough with existing content to fit a recognizable pattern of expectations for user engagement; that is, the audience must understand how they are expected to engage with new content based on their previous experiences with the interface media.

When creating new content for use in a CMS, students should reflect on the needs and values of both stakeholders and various audiences while also optimizing that content for delivery using the affordances of media available in the CMS. And students should account for the rhetorical velocity and spread of content into different channels. Content created with any CMS in mind should draw on/complement the strengths of that system, supplementing them when necessary to increase accessibility for potential stakeholders.

This combination of factors that influence the distribution of CMS content means that students must draw on the existing strengths of the system to create modular content that might be used in a variety of media to achieve a variety of purposes in respect to various audiences. These content pieces should be small—fractions of individual posts or documents—and they should be designed to utilize the various affordances of the CMS media.

Step 6: Use Content Creation to Compensate for Shortcomings in the System

This heuristic assumes in part that the CMS and its interface with which the students are working have the potential to adequately meet the needs of the target audience most of the time. However, there will inevitably be situations for which the CMS is not optimized. The students must have a thorough understanding of the needs and expectations of the audiences and stakeholders to recognize these situations when they arise and make adjustments to the content they create to help compensate for any shortcomings.

Case Scenario

Instructor Alex Garden has been teaching an introductory course in technical and professional writing for many years but has grown increasingly uncomfortable with the number of single-author assignments that students complete. The emphasis on stand-alone documents and genres does not account for the changing ways technology has influenced the writing and composing process. She knows that in most modern workplaces and contexts, technical writers work collaboratively on documents, often composing brief sections of content and pulling small pieces of existing content together to be used in a variety of media. She decides that students need practical experience with this kind of writing as well as a theoretical understanding of how such writing relies on a series of other components. She imagines a series of projects or even a course where students work with a hypothetical client, like their university department, who needs to communicate similar information with a variety of audiences and stakeholders through a number of media outlets on a regular basis. The university is especially interested in communicating some of its inclusive practices when it comes to supporting students, so inclusive audience analysis will be important to such a course or unit.

To begin the project, Alex Garden asks students to examine the media outlets that their department uses and compare these with similar departments in peer institutions in their region of the country. They find that their department, like others of its kind, is well represented on social media sites including Facebook, Instagram, and Twitter. It also has an extensive university department website, operated through *WordPress*, and a number of videos, pamphlets, fliers, and informational brochures are delivered through this CMS throughout the year. Alex Garden then asks students to think about and answer the following questions: why does the university need so many media outlets? Why do these outlets, in particular, make sense? To whom does each form of media appeal, and why? What kinds of content are found within these digital spaces?

Step 1: Analyze the Active Interfaces Maintained by the CMS and Complete a Content Audit

Alex Garden's class completed its thorough review of existing media for the department. Now, in order to write manageable content to be used across these media consistent with their affordances and the expectations of users, the students need to understand how these media are currently being used and how they have been used in the past. They begin by completing content audits of the department website, major social media accounts, and the printed informational documents available in the departmental office.

Step 2: Understand the Affordances, Constraints, and Conventions of the CMS

Students in the class now need to analyze the affordances, constraints, and conventions of the social media used within the CMS to see how these align with the content that has historically been produced. What specific types of content can be delivered using Twitter, Facebook, and Instagram? Is there a limit on the number of characters? Can photos or images be shared? Which platforms use hashtags or tagging so that the content can be shared? Answers to these questions will give students insight into the types of content shared and distributed through these social media channels. This analysis provides students with opportunities to understand the velocity and spread of specific content throughout the ecology.

Step 3: Determine the Needs and Motivations of Audiences and Stakeholders

Alex Garden's class now knows that its department communicates about policies, upcoming events, campus news, and academic achievements through the department website, major social media accounts, and a combination of fliers, pamphlets, and brochures. It also understands the affordances, constraints, and conventions of these media. Now, Alex Garden needs to get students thinking about the audiences who will eventually use (and even distribute) their content. She begins by asking students to brainstorm who might use each medium, and why. The students note that they often use the department website themselves to look at degree requirements and course offerings; a few note that they follow one of the social media accounts, and some mention that their parents follow the department on Facebook. The class then notes faculty and staff, administrators, potential students, alumni, and funding sources as potential audiences. Alex Garden asks students to interview a range of users they know who are a part of these audiences and determine their motivations for engaging with these media. Together, the class makes a chart of potential audiences, their motivations, and their needs in order to articulate the primary audiences for the content they create.

Step 4: Determine How the Active Interfaces Work to Meet the Needs and Motivations of Audiences and Stakeholders

The students have already identified the affordances, constraints, and conventions of the CMS and CMS interface media; they know the kind of content that is typically included in this ecology, and what the needs and expectations of the audiences and stakeholders are. This fourth step is an act of synthesis, identifying how and why all of these features are able to co-adapt to create successful systems. Alex Garden helps students with this difficult process by asking them to draw parallels among the needs and expectations of the various audiences and the features that help meet these needs and expectations. The students are likely to find this process intuitive, but they may struggle with articulating the reasons why these connections work. However, we find this articulation focus helpful because once students understand why the systems work, they are better able to make necessary adaptations without disrupting the system's success.

Step 5: Create Content Based on Audience and Stakeholder Needs and Values within This System

Alex Garden's class is ready to start creating content. She decides to have students start with designing content about a new minor that can be distributed through a variety of media within the CMS. The students agree that one thing the audiences and stakeholders will need from this content is a list of courses required to complete this minor. They consider that such a list—one that contains all the required courses—will be most helpful for audiences thinking about the minor as a long-term, coherent unit, so they create a textual list that gives the titles of each course in the order in which the courses should be taken. They also think that course descriptions will be useful, but recognize that the length of these descriptions will vary depending on the medium of distribution and other content in that document, so they create one-sentence, two-sentence, and paragraph-length descriptions. One student points out that many of the CMS media are multimodal, so they also create course descriptions in the form of video clips and 35- to 45-second video files. All these different iterations use similar keywords and phrasing, but these content pieces can now be used in a variety of contexts within the CMS, depending on the needs and expectations of the audience.

Step 6: Use Content Creation to Compensate for Shortcomings in the System

Alex Garden's class has been creating outstanding content, but several students have begun to notice a problem: their university and department has been explicit about prioritizing diversity and inclusion in campus initiatives, but most of the media associated with the CMS is steeped in discursive practices that privilege access for individuals who have the social and financial freedom to spend a great deal of time on campus. Even the social media pages tend to rely on networks of personal interactions formed in this way, meaning that many students and their families are left out of departmental communications. From their earlier research, the students know that Twitter places more emphasis on hashtag topic organization, and less on externally established networks to help users connect. For this reason, they believe that directing interactions to Twitter will help the department content they are distributing to become more accessible to more people. They begin to include specific references to relevant hashtags on content distributed via other media, hoping to drive conversations onto this platform.

Conclusion

We conclude this chapter with two brief scenarios that instructors can use in courses to have students practice creating manageable content that is inclusive of their audiences. Using the heuristic described in this chapter, students can engage with their local universities or communities in authentic ways. Teaching students to create manageable, inclusive content depends on having CMSs and authentic contexts where students can practice and then analyze how their content is received by their audiences.

Scenario 1

Your class has been asked by a local fitness center near the hospital for help in promoting physical activity with an adult population that have historically had health issues. You are tasked with persuading people who are not at all active to increase their physical activity. The idea is to promote physical activity in a wide variety of forms, but losing weight should not be the primary benefit or be mentioned in any of the marketing materials. The content created needs to be created and designed to be used in the following contexts: fliers, brochures, postcards, website, Facebook, Twitter, Instagram, etc. Have students use the heuristic introduced above to generate content for this fitness center.

Scenario 2

Your class has been contacted by a local professional organization known for its role in facilitating conversations about science and technology in society. Throughout its existence, the organization has struggled to attract new members; recently, this phenomenon threatened the continued existence of the organization. Your class has been tasked with creating content items to advertise the organization and increase membership and participation. These content objects will be shared across a variety of digital and nondigital platforms and are intended to reach a variety of audiences, including current local university students in fields related to science and technology in society, recent graduates/graduating students in these fields from universities around the country, international recent graduates in STEM and humanities fields, local professionals interested in networking in their field, and STEM-focused corporations interested in ethical business practices and promotion of their products. Have students use the heuristic to help this organization meet its goals.

References

Batova, T., & Andersen, R. (2016). Introduction to the special issue: Content strategy—A unifying vision." *IEEE Transactions on Professional Communication*, *59*(1), 2–6. doi:10.1109/TPC.2016.2521921.

Batova, T., & Andersen, R. (2017). A systematic literature review of changes in roles/skills in component content management environments and implications for education. *Technical Communication Quarterly*, *26*(2), 173–200. doi:10.1080/10572252.2017.1287958.

Clark, D. (2008). Content management and the separation of presentation and content. *Technical Communication Quarterly*, *17*(1), 35–60. doi:1080/10572250701588624.

Clark, D. (2016). Content strategy: An integrative literature review. *IEEE Transaction on Professional Communication*, *59*(1), 7–23. doi:10.1109/TPC.2016.2537080.

Davis, C. (2018, February 11). *Memetic rhetorical theory in technical communication: Re-constructing ethos in the post-fact era* (Doctoral Dissertation, East Carolina University). Retrieved from http://hdl.handle.net/10342/6935.

Gonzales, L. (2018). *Sites of translation: What multilinguals can teach us about digital writing and rhetoric*. Ann Arbor, MI: University of Michigan Press.

Gonzales, L., Potts, L., Hart-Davidson, B., & McLeod, M. (2016). Revising a content-management course for a content strategy world. *IEEE Transaction on Professional Communication*, *59*(1), 56–67. doi:10.1109/TPC.2016.2537098.

Haas, A., & Eble, M.E. (2018). *Key theoretical frameworks: Teaching technical communication in the 21st Century*. Logan, UT: Utah State University Press.

Hart-Davidson, W., Bernhardt, G., McLeod, M., Rife, M., & Grabill, J.T. (2007). Coming to content management: Inventing infrastructure for organizational knowledge work. *Technical Communication Quarterly*, *17*(1), 10–34.

Johnson-Sheehan, R. (2005). *Technical communication today*. Pearson/Longman.

Ridolfo, J., & DeVoss, D. (2009). Composing for recomposition: Rhetorical velocity and delivery. *Kairos: A Journal of Rhetoric, Technology, & Pedagogy, 13*. Retrieved from http://kairos.technorhetoric.net/13.2/topoi/ridolfo_devoss/.

Rose, E.J., Edenfield, A., Walton, R., Gonzales, L., Shivers, A., McNair, T.Z., … Moore, A. (2018). Social justice in UX: Centering marginalized users. In *Proceedings of the 36th ACM International Conference on the Design of Communication* (SIGDOC '18), 2 pages.

Sun, H. (2012). *Cross-cultural technology design: Creating culture-sensitive technology for local users*. New York, NY: Oxford University Press.

9

WRITING ABOUT STRUCTURE IN DITA

Jason Swarts

NORTH CAROLINA STATE UNIVERSITY

Chapter Takeaways

- An understanding of the rhetorical challenges of writing with structured and componentized content.
- A structured sequence of lessons for teaching basic concepts and practices of structured authoring in the Darwin Information Typing Architecture (DITA).
- Writing and assignments and resources on DITA and topic-based authoring that are appropriate for writers who are new to DITA technology and topic-based writing.

Decades of research on the job market for technical communication have revealed the extent to which communication and technological proficiencies have become entangled (e.g., Lanier, 2009; Blythe, Lauer, & Curran, 2014). Since the mid 2000s, those proficiencies have started to include writing with content management systems (CMSs). To work within those authoring environments, writers must be able to work with "more varied—and more nimble—forms of content" than they might have been trained to use (Brumberger & Lauer, 2015, p. 234). However, technical communication programs still largely focus on generating documents while the technical communication industry is increasingly interested in generating content (2015, p. 239). Consequently, employers are finding that graduates of technical and professional communication programs often "have had little exposure to the processes that underlie modern content development and knowledge management" (Doherty, 2017, p. 1).

A number of challenges come with teaching writing for a CMS environment that is modular and structured. One challenge is assessment-based: how does one assess student writing if the documents produced are assembled from granular bits of content, some of which might be collaboratively written, pre-written, and reused? Another challenge is tools-based, for the tools used to produce structured and modular writing can be perplexing. Yet it is through engagement with these tools that students will develop the conceptual understanding of working with content and content structures.

One possibility for bridging this gulf of experience is to teach students a form of markup for creating structured, modular content. One such protocol is DITA, which was developed by IBM in the early 2000s (see Priestley, 2001; Evia & Priestley, 2016) and is currently a protocol of choice for many organizations worldwide. In recent years, the protocols and markup behind DITA (and similar protocols) have been incorporated into writing technologies like *Adobe FrameMaker* and *Oxygen XML Editor*, which flatten out the learning curve by putting the markup in a more familiar word-processing interface. Yet, as familiar as the application of DITA might become through a word-processing interface, some underlying concepts related to modular writing can challenge a student's sense of text and writing.

Each semester, in a graduate class on technical writing, and on a smaller scale in an undergraduate class on writing software documentation, I ask students to collaborate on documenting a software package with modular, reusable chunks of content that can be structured together into multiple documentation outputs through a CMS. This assignment brings us face to face with technological and rhetorical challenges that technical communicators must recognize and face to develop the conceptual understanding of writing for content management that employers are seeking. Students learn how to think about documents in terms of structures; they also learn DITA authoring tools, and reflect on the affordances and constraints of those authoring tools. Their conceptual understanding then consists of understanding text structures, modular content, and reuse. They also gain insight about the rhetorical characteristics of DITA-structured writing, characteristics not readily apparent (see Clark, 2008). I discuss two characteristics: bilocational writing and translocational writing. I first describe the importance of DITA-structured authoring in technical communication and then, through the lens of my experience, discuss how the reader might teach these concepts to students of technical communication.[1]

What Is DITA?

Although the context of this chapter so far has been technical communication, it is not just technical documents that have a discernible and conventional structure. Many documents have a structure to them, and the more

1 This chapter is not the place for a full exploration of DITA, but there are many good resources on the web, such as learningDITA.com and www.scriptorium.com/learning-dita/.

common and typified the document is, the more likely it is that readers will recognize the structure. Documents like encyclopedia entries, prescription information, and telephone books have readily identifiable structures that facilitate use. However, other documents, like five-paragraph essays, "bad news" letters, and proposals also have discernible structure as well, albeit with more variation.

Keeping with technical communication, one of the most iconic forms of structured writing is the documentation set. For nearly the last four decades, the structure of technical documentation has remained relatively stable (Van der Meij, Karreman, & Steehouder, 2009). The structure often consists of a title that designates an activity, a conceptual element that explains background information or defines key terms, infinitive subheadings that divide an activity into discrete tasks, steps that break down a task into discrete actions, and notes that comment on those steps (Farkas, 1999, p. 44). The typified components of the documentation structure help readers to (1) identify a desired state, (2) acquire prerequisite information, (3) monitor progress through steps toward an intended outcome, and (4) anticipate and recover from errors (Van der Meij & Gellevij, 2004, p. 5).

This archetypical documentation structure is at the heart of structured writing with DITA. In early iterations of the DITA protocol, the basic structural elements of the markup system were *concepts*, *tasks*, and *references* (Priestley, 2001). Concepts captured information regarding the desired and prerequisite states. Tasks captured information related to the steps and monitoring of progress. The references captured information that users might need as they move through the tasks (Bellamy, Carey, & Schlotfeldt, 2011, p. 11).

Beyond reinforcing the structure of technical documentation as a whole, DITA also applies structuring to the contents of concepts, tasks, and references. Concepts, tasks, and references are separate files in DITA, "topics" that follow their own set of structural rules. For example, the following excerpt shows a task topic that might consist of a title, a step, an illustration, an example, and a note.

```
<task>
  <title> Creating a Bibliography </title>
  <shortdesc> Easy Cite can create a bibliography of your
  cited sources from the entries contained in its data-
  base. </shortdesc>
<taskbody>
  <context> To create a bibliography from your cited
  sources: </context>
  <steps>
        <step>
              <cmd> Right-click an entry to add it to
              your bibliography. </cmd>
```

```
            <info> Hold down the Shift key to select
            more than one source. </info>
        </step>
       [...]
  </steps>
</taskbody>
</task>
```

Each topic has its own set of structural elements and rules for arrangement. For example, a <step> must contain a command (<cmd>). Instead of tasking writers with the responsibility of visually shaping the appearance of their content, DITA requires writers to apply structure to give that content semantic value (i.e., what kind of content is it). The structural elements have their own rules for appearance. While on one hand the adoption of the DITA markup makes writing more cumbersome, a distinct advantage is that the structured approach allows for a clean separation of content from its appearance that better allows "content reuse and multichannel publishing" (Andersen & Batova, 2015, p. 256).

When DITA-structured content is saved as componentized bits of structured content in a CMS, that content can potentially be pulled into a number of publications that might use the same information (e.g., tasks and concepts) in a document with a different overall structure (e.g., release notes).

Why Technical Communicators Use DITA

Without meaning to sound cynical, one prime motivation for adopting structured writing is that it makes writing, collaboration, translation, and publication of content more efficient and cost effective (Andersen, 2013). Beyond the business incentives, structured writing has other affordances, such as allowing adaptation of content across different genres that rely on similar kinds of content, and maintaining visual consistency across documents (Clark, 2008, pp. 54–55).

The tools for marking up content in DITA simplify the process of content creation by offering premade content tags and rules for assembling those tags, which in turn assist in standardizing texts, and allowing authors to use that content in different outputs (Evia & Priestley, 2016, p. 24). These patterns of reuse are essential to organizational-level content management (see Rockley, Kostur, & Manning, 2003), and it is no surprise that tools for producing such content are becoming easier to use. Learning the tools, however, is not the only objective when training technical communicators.

Writing is obviously more than the mastery of tools. Skilled writers understand user experience and how readers interact with their texts and use them toward foreseeable ends. DITA-structured writing also affords writers the ability to apply what they know about readers and tasks. The structures discipline writers to produce the content and to rely on the structured outputs to help readers "find the information they need faster, accomplish their goals more efficiently, [and] read only the information they need to read" (Bellamy et al., 2011, p. 17).

DITA-based writing also facilitates new user experiences, allowing them ways to access and use information differently (Clark, 2008, p. 56). And learning about these new user experiences can be just as critical to our understanding of the tools we are trying to use. Two rhetorical characteristics I have talked about with students help them see that how I address readers' concerns depends on how readers experience the content in a DITA-structured document. Bilocational meaning is that that arises from interpreting content in multiple contexts simultaneously. Translocational meaning is that that arises from interpreting content across contexts.

Bilocational and translocational meaning is largely afforded by the dictum in DITA authoring to produce content chunks free enough from context that they can be reused. Of course reuse is valuable from a single-sourcing standpoint, where content can be written once and used many times, but another advantage is that topics written for reuse can be wired together in ways that create new texts, depending on the paths that readers take through them. Still, writing for reuse is challenging because of the inherent intertextuality of language (see Bakhtin, 1981) and our unconscious tendency to use relational and functional language (e.g., pronouns, indexicals, and relative modifiers) that point to a context.

Bilocational

One affordance of DITA-based authoring is that information can be in more than one place at one time but also in multiple places at the same time, serving different functions. For example, the short description <shortdesc> element serves both as meta information on a topic, and a first paragraph in the topic, as well as the content one would see in a search preview. The content reference <conref> element allows granular bits of content to be reused across topics, bits as small as product titles and informational notes, but also larger pieces of content. Writing for bilocational meaning is possible because of the efficiencies of reuse and revision afforded by DITA, but the rhetorical affordance of bilocation goes beyond the practicalities of content reuse to a doubling or multiplication of rhetorical effect.

Bilocated content bears the traces of multiple meanings, audiences, and invoked contexts of use and interpretation (see Swarts, 2010) and in some contexts of interpretation content known to appear in more than one place acquires a unique depth that can link readers together (see Schryer, Afros, Mian, Spafford, & Lingard, 2009). Authors are developing techniques that help them both understand the material they are drawing upon and manage the different meanings in their compositions (see Leijten, Van Waes, Schriver, & Hayes, 2014).

Translocational

Another affordance of DITA-structured writing is that the information in a text is more modular and componentized and wired to other chunks of content that create non-obligatory reading paths. The topics that comprise a piece of documentation are linked together with maps and submaps that act as superstructures to present the content in a genred form while also allowing readers to choose flexible paths for experiencing the text. Although the DITA project might have a map that defines a hierarchical set of relationships between topics and reinforces a particular structure to the information, readers can also pursue lateral paths of investigation. Meaning arises from the user's movement through the topics that comprise the outputted document.

To work with translocational meaning, writers of DITA-structured texts must consider how to address user expectations and create a reading experience that allows those users to understand what they have read and how to apply it to their particular ends. One approach to building a sense of translocational meaning is for writers to begin with an audit of the content that one has for documentation (see Batova, Andersen, Evia, Sharp, & Stewart, 2016) to determine whether that content is written to address the intended audience at the right level of experience (Andrade & Novick, 2008). The difficulty with this approach is that we leave untouched the complicated issue of understanding tasks. We know that user tasks are complex (Mirel, 1998) and that it is difficult to structure documentation to reflect that complexity. DITA-structured content has the advantage of being flexible, which allows us to come back to the root of this problem; that is, that the very idea of a task is situated in a way best understood by users (see Kohlhase & Kohlhase, 2009, p. 136).

DITA-structured content has a suggested structure to it, but no obligatory one. Users can access content by following a train of thought or by following an "information scent" (Morville, 2005, p. 60) that might be picked up by finding content on one topic and following it to related topics. The resulting movement across locations (i.e., translocation) is the process by which readers apply a frame derived from their emerging understanding of their task to make sense of the information they are accessing.

Although writers are developing techniques for assembling documents from reusable, structured content, one of the remaining challenges is to help writers get a "sense of text" that knowledge of "the structural and semantic arrangement of the text—the absolute and relative location of each topic and the amount of space devoted to each" (Hansen & Haas, 1988). Hansen and Haas may have been concerned with the transition from handwriting to computer-composition, but the same disorientation can be expected when working with DITA-structured content. A noted usability challenge of working with structured content is that to make reusable content and gain the benefits of single sourcing, writers need to create content capable of being used in different settings without foregoing its ability to be useful and applicable in any one of those of those settings (Sapienza, 2004; Andersen & Batova, 2015, p. 254).

Teaching DITA-structured Writing

The approach I recommend for teaching DITA-structured authoring is project-based. Because the advantages of CMSs and DITA are best seen in somewhat complicated and large documentation projects, those kinds of assignments will highlight the practical advantages of DITA-structured authoring while also getting students to grapple with the challenges. What I mean is that in a large-enough project, students begin to understand what is advantageous about writing content that can be reused. They also come to appreciate the difficulty of keeping track of multiple concept, task, and reference topics and smaller pieces of content that support bilocational and translocational uses among readers.

Going into the course, one should assume that the students know nothing about DITA and basically nothing about structured authoring. Some students may have some understanding of HTML, and that is a useful starting point for talking about markup and adding structural information to content.

As a preliminary to the more DITA-specific instruction that I discuss in the coming sections, I recommend that you work with students as you would when planning any large documentation project. Help students learn about audience and task analysis and engage in primary research to learn about the audiences that they will be trying to reach. Based on this information, the students will be able to start understanding what some of the concepts and tasks are that should go into their documentation. Then do a content analysis of any instructional content available elsewhere.

With a documentation project in mind, the class is ready to learn basic elements of DITA, starting with the idea of structure and then basic DITA topic types (e.g., concepts, tasks, and references). At this point, students will be ready to learn more about the content types that make up the basic topics, and they will be ready to learn how to write that content for maximum reuse. As soon as the students start to consider how to write for reuse and how to shape the reader's experience of the topics and contents, they will be ready to confront the challenges of writing for bilocational and translocational meaning.

Teaching Structure

As mentioned earlier, documentation in technical communication has a definite structure to it that most of us recognize when we see it, even if we cannot always neatly identify the different component pieces. Instead of leaping into the structure of computer documentation, build up to that example by looking at simpler structures.

Assignment: Look at four different documents, including a business card, a recipe, a resume, and an introduction to an academic paper. It may be helpful to look at multiple examples of these document types if any of them are unfamiliar.

- Identify common information types appearing in the document.
- Determine the relationship between pieces of information (e.g., dependencies, order).
- Determine the rules by which the information types are ordered, yielding the document in question.

Identify the information types and give them descriptive names. Looking at the placement of information on the page and using any clues available from the layout (e.g., indentation) determine if there are any relationships between those pieces of information and rules that determine how the information is laid out on the document.

The result of an assignment like this is that you will have a set of information types that can be converted into XML, which are essentially user-defined semantic tags. Getting the syntax right at this stage is less essential than getting a feel for the idea of a document structure. For example:

```
<recipe>
<classification>
   <category>dessert</category>
   <title>Gelatin Mold</title>
   <servings>4</servings>
<classification>
<ingredientlist>
   <ingredient><amount>1 pkg (6 oz.)</amount> lime gel-
   atin</ingredient>
   <ingredient><amount>1 cup</amount> water</ingredient>
   <ingredient><amount>8 oz</amount> cottage cheese</
   ingredient>
   <ingredient><amount>1 cup</amount> walnuts <pre-
   p>chopped</prep></ingredient>
[...]
</ingredientlist>
<directions>
<step> mix gelatin into boiling water [...]</step>
[...]
</directions>
</recipe>
```

Have students try the same approach on the other texts, including resumes and introductions to academic articles. They will quickly see how the exercise becomes more difficult as the texts become more complex and variegated. The advantage of using this exercise with more complex document types is that students will begin to see how pieces of content that they may think about as conceptually distinct (e.g., a claim, a warrant, and evidence) are difficult to isolate in a text, either because those pieces of content are not contiguously presented or because they contain intertextual references to other pieces of content. Seeing this mixing together of content types will set students up to appreciate how difficult it can be to write content that can be reused.

You can move from an assignment like this one to technical documentation. Sources like Farkas (1999) and Van der Meij and Gellevij (2004) are particularly helpful in pointing to standard information types and conventional ways of structuring those information types in documentation. This layered approach to understanding the abstractions associated with structured writing is one that others have recommended as well (Evia, Sharp, & Pérez-Quiñones, 2015).

Take this opportunity to begin teaching students about the basic topic types in DITA: concepts, tasks, and references (see Priestley, 2001; Bellamy et al., 2011). They are likely to understand these topic types after having familiarized themselves with conventional forms of technical documentation. And with this knowledge, help students to start thinking through their documentation project, assigning pieces of content to particular topic types. What are the conceptual elements and what tasks follow from them? What reference information will users require?

With the list of concepts, tasks, and references in mind, a good next step is to look at the makeup of these DITA topics. Although it is possible to begin writing directly in a WSYWIG XML authoring program on a DITA template (e.g., *Oxygen XML Editor*), a more helpful approach, from an invention standpoint, would be to look at the structure of DITA topics. A visual summary of DITA concept, task, and reference topics appears in the Appendix to this chapter. You can use these visualizations to teach students the structural relationships between elements of content that appear in those topic types.

The value of this and the other structure diagrams is that they show the order and relationships between different kinds of information in a topic. DITA elements that sit on the same row have familial relationships. The DITA elements to the left are parent elements of the child elements on the right. Therefore, a student can see that a task topic must have a <taskbody>, and within that are <steps> and commands <cmd> that present steps in their common and recognizable form. The students will also see the relationships between notes, information, and warnings and the steps to that they refer. They can also see suggestions for content that might be useful to create, such as step examples (stepexmp). Each DITA element strongly suggests to the students what the topic of a task looks like, what kind of content it could have, and what pieces of content are understood to be semantically different from each other. For example, knowing that a step example <stepxmp> is separate from the step/command itself <cmd>, but that it exists at the same level of importance, is helpful to understand when writing them. Commands should not contain their illustrative examples. The reason is that one might reuse the command or the step example in another setting.

The diagrams convey a sense of structure while also providing a reduced list of elements that students can use to begin inventing content (see a related approach in Evia, 2019). These structures also clearly communicate the kinds of constraints that writers work under (e.g., content dependencies). In other words, the structure diagrams provide a sense of the text, a picture of the genre of documentation as seen through the lens of DITA. Learning DITA from this point is a matter of learning specifics about the tags and practicing their use.

Teaching Bilocational Meaning through Reuse

Once students start writing their topics, following the structure diagrams supplied here, they are going to recognize that there are some elements and pieces of content that appear in more than one place.

The first such piece of content is the <shortdesc> element, which is used to introduce a topic by describing the following components:

- An overview of the procedure.
- The benefits or importance of the task.
- Limitations or requirements.
- Brief conceptual information (Bellamy et al., 2011, p. 25).

The <shortdesc> does more. It is a link preview, meaning that when someone searches for documentation, the <shortdesc> is the preview text returned with the link. For this reason, ask students to consider how the <shortdesc> both introduces a topic and provides a meaningful overview of the contents while also standing alone as a statement that entices viewers to follow a link. The <shortdesc> cannot refer to the steps below it and probably ought not to include too much contextual information that might seem out of place in a set of search results. The bilocational function of the <shortdesc> makes it one of the more versatile information components in a topic and one of the more difficult to write.

Just as the <shortdesc> can have bilocational value, so too can any piece of content in a DITA-structured set of documentation, and this is where one of the biggest values of writing reusable content becomes apparent. Any piece of content, a phrase, a sentence, a paragraph, an image, can be reused in other topics. The challenge to reuse is the same as what applies for the <shortdesc>, but unlike the short description, which is a stand-alone paragraph, other reused content can come directly from other content structures: an example would be a step or an example within a set of steps. And so, an important lesson and assignment to give to students would be to recognize how to write content in a way that reduces intertextual references and other language that situates it within a context that might not be pertinent to all users in all contexts of use. Sometimes, that content is written in a way noticeably situated in the context for which it was originally written.

> **Assignment**: Learn to write context-free content. Choose an example of documentation from any source and choose a topic containing content that might be potentially reused elsewhere in the documentation. Edit that fragment of content to remove contextual information that situates it to a particular location in the text or to a particular point in time. Keep in mind four basic guidelines for writing content that can be reused:
>
> - Write topics that'll make sense when taken out of context.
> - [I]nclude organizational features that make the content easy to scan and easy to carve out for reuse elsewhere.
> - Don't use relational language.
> - 'Write componentized content that is feature-based so that it can be reused.'
>
> *(Bellamy et al., 2011, p. 196)*

Ask students to find a piece of potentially reusable content and revise that content so that it can be reused. In the process, they should remove content referring to specific versions of the software, or any language that points to other content or images on the page around it. For example, "Before making any changes to content in *Moodle* 3.3.1, editing must be turned on." turns into "Turn editing 'on' before making changes."

In this case, the phrasing is revised into an imperative form. The reference to *Moodle* 3.3.1 is cut, and the phrasing now reflects the name of the interface element to be selected, "Turn editing 'on.'" By comparison to the original, the revised phrase is more reusable and can serve a rhetorical purpose in multiple locations: as a step in a task, a note in a reference topic, or a contextual element in a concept topic. The content still retains its form as an imperative and so it is most likely to be reused as a command of some sort, but the stripped-down form of the content allows it to travel across contexts. Students will be able to add context back into these topics, but at the level of metadata (see the next section on teaching toward translocational uses of structured content).

Part of the challenge for students will be to envision the different kinds of rhetorical work that reused content must do and pick up from different topical contexts. Another challenge will be to isolate the content worth reusing and flag it with memorable IDs. Continuing with the running example, the content "Turn editing on' before making changes" might be found in a topic called "c_new_course.dita" where the content would be flagged with an ID:

```
<concept id=new_course>
<title> Building a New Course</title>
[...]
     <conbody>
     [...]
          <cmd ID="editing_on"> Turn editing on before
          making changes.</cmd>
     [...]
     </conbody>
</concept>
```

In the target document, where the content is to be reused, the tagged ID would be called into the new document with some indication of intended use:

```
<task id=adding_video>
<title> Adding Video </title>
[...]
     <taskbody>
          <steps>
               <step><cmd   conref="Concepts/c_new_
               course.dita#new_course/editing_on"/>
               </step>
     [...]
          </steps>
     </taskbody>
</task>
```

In this example, the code inserts a <cmd> element and draws the content from the file "new_course.dita," which is in the "Concepts" folder. Within the "new_course.dita" file, that has a topic ID of "new_course," DITA will look for an element ID called "editing_on" and then mirror that content.

Using whole paragraphs and topics might be some of the most common ways that content is reused in DITA, but it is possible to reuse content at the phrasal level, which allows writers to convey the same content consistently, forestalling errors in interpretation and translation.

The instructional aim of teaching content reuse for bilocational reading practices would be to get students to think about what content should be reused and in what topical or output settings. When students have a sense of what content should be reused, they should make sure that the pieces IDed for reuse will work cohesively with the other content in the target document and not point to other tasks, concepts, or references that have no relation to the target topic.

Teaching Translocational Meaning through Modeling, Mapping, and Metadata

After learning some fundamental DITA topics, students can then turn to the tricky issue of structuring topics to afford a translocational experience. This process begins by returning to audience and task analysis.

> **Assignment**: Review the kinds of tasks that users bring to the technology. What concepts and tasks and references do members of your intended audience associate with one another? Are there groupings of tasks that would suggest a common sequence of tasks? What are the terms that your readers use when referring to these concepts, tasks, and references?

What students will discover from their interviews with users is that tasks do not always neatly overlap with tools or functions in the software. A single task might involve using multiple tools in sequence or in parallel. Ask students to identify these task trajectories, visually and topically, to discover that topics (i.e., concepts, tasks, and references) are functionally related. Students will discover that those interface tools and information related to those tools are part of their view of the task (see Hart-Davidson et al., 2005).

Knowledge of how users experience a technology and think through their tasks helps writers with two important steps in planning DITA-structured projects, both of which contribute to the user's translocational experience. The first is that the interview data will help writers discover what paths readers take between topics. This translocational movement through different topics reveals the kind of context readers find meaningful for interpreting a task. What are the concepts, tasks, and references that users find valuable? Furthermore, how are those topics related to one another and how does one suggest or point to another? What tasks belong to what concepts? What references belong to what tasks? Use the results of this kind of inquiry to create a map that shows the relationships between topics (see Figure 9.1).

These topics are locations or spots in the task that guide the users. They also constitute an "information model" that can guide production of the topic maps (Bellamy et al., 2011, p. 96). The information model will also be

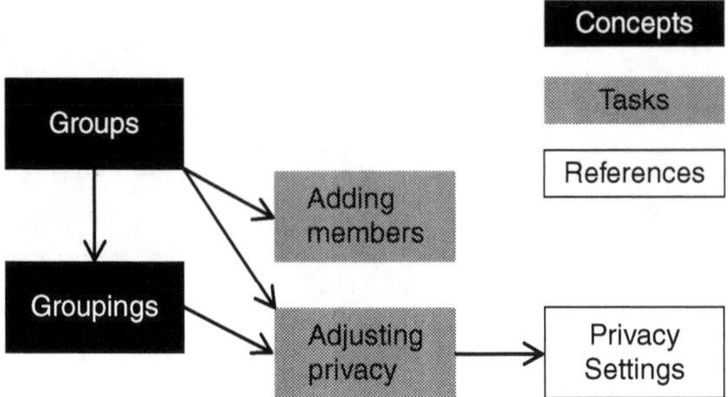

FIGURE 9.1 Information map (partial)

valuable when structuring the overall document. This structure, the relationship between the topics, will be captured in a ditamap that shows an intended relationship between topics. It will also be captured in a table of contents through the use of <topicref> or topic reference elements that call the files holding each discrete topic and place them into a hierarchical relationship with other topics, as in the following example:

```
<map>
    [...]
    <topicref href="Concepts/c_groups.dita">Groups
         <topicref href="Concepts/c_groupings.dita">
         Groupings</topicref>
         <topicref href="Tasks/t_adding_members.
         dita">Adding Members</topicref>
         <topicref href="Tasks/t_adjusting_privacy.
         dita">Adjusting Privacy
         <topicref href="References/r_privacy_settings.
         dita">Privacy Settings<topicref>
         </topicref>
    </topicref>
    [...]
</map>
```

A translocational experience of a task, however, need not be shaped just by the structure of the ditamap. As users proceed through a task, different related tasks or parallel and simultaneous tasks might also be occurring. And accommodating this view of tasks is a distinct affordance of DITA structuring. Of course, writers can create links between topics and create a networked organizational structure within a topic, but that is not always a recommended way to proceed. A different approach that allows users to build routes through or

TABLE 9.1 A Relationship Table

Concept	*Task*	*Reference*
Group	Adding Members	
Grouping	Adjusting Privacy	Privacy Settings

views of the information in this document is to build a series of related links. Writers facilitate this translocational experience by drawing those topics close to the surface, making them visible, but also by controlling the nature of the relationship. Ask them to create a table showing those topical leads a reader might follow (see Table 9.1).

Table 9.1 is a simple example of a concept called a relationship table (<reltable>) that appears in the ditamap and allows writers to show "related topics" and to control the reader's flows through those topics. At all times, the potential relationship between topics is apparent but not necessary: the <reltable> shows each rhetorical location invoked by the tasks learned about in the user interviews and present here to assist users.

Still further control over this translocational experience is possible when writers control the direction of linkage through the topics referenced in the <reltable>. Linkages can be one- or two-way links. Writers can designate whether a topic can be both the source of a link (i.e., where the relationship originates) or the target of a link (i.e., where the relationship goes). A "source only" link designates a topic that can only be the source of the link. A "target only" link can only be the target of a link. For example:

```
<reltable>
   <relheader>
      <relcolspec type="concept"/>
      <relcolspec type="task"/>
      <relcolspec type="reference"/>
   </relheader>
   [...]
<relrow>
      <relcell>
      <topicref href="Concepts/c_groupings.dita"/>
      </relcell>
      <relcell>
        <topicref href="Tasks/t_adjusting privacy.dita"/>
      </relcell>
      <relcell>
```

```
        <topicref href="References/r_privacy_settings.
      dita" linking="targetonly"/>
      </relcell>
    </relrow>
    [...]
</reltable>
```

An example might be a link from a command in a task to a supporting reference topic. However, if that reference topic is used in a variety of tasks, it may not make sense to link from that reference to any one particular task topic. Although the relationships in the <reltable> may be built from the interviews that students have with potential users, they may need to be adjusted or changed over time, as new topics are added and existing topics are changed. The point, however, is that a user's task sense can be portrayed as a translocational experience of the full documentation set, bringing different componentized pieces of DITA-structured text to the surface in a way that would not be as easily accomplished or maintained without CMSs.

Finally, one additional lesson to teach in support of users' translocational experiences in DITA documentation is on the use of metadata for adding keywords to topics both at the level of the ditamap and inside the topics themselves. Users navigating through a set of topics might not find the hierarchical structure of the ditamap to be helpful, nor the relative linking between the topics. However, the audience- and task-based interviews are as likely as not to produce a set of terms and nested concepts that could support search within or across topics.

Inside the <prolog> element of any concept, task, or reference, writers can add <indexterm> or <keywords> and also embed keywords inside one another to support a view of looking at a topic that would suggest what the topic is about primarily and what its secondary meaning might be. Although searching might be a last-resort kind of navigation for a collection of documentation, users' translocational experiences with the content might make search the most viable option for finding relevant topics. Help students come up with a good keyword system to support such user access.

Conclusion: Writing with Structure

Despite the widespread use of DITA, Markdown, and other kinds of structural markup for documentation, the tools are not without their critics. The principal complaint about DITA-structured authoring is that reducing the documentation process to the mastery of DITA elements and topic forms strips some of the creativity and rhetorical decision-making from the process of writing. Writers make fewer decisions about the kind of content included in documentation, and fewer decisions about how that content should look, and in the case of writing for reuse, perhaps develop less of a sense of text when writing context-less, componentized content. A more generous view of DITA-structured authoring is that while the technology can hamper some elements of creativity, it can also open up new possibilities for rhetorical expression, for writing content that can be assembled into new meaningful forms (see Clark, 2008; Evia & Priestley, 2016).

The points that I have made about understanding structure and writing reusable content are a way of identifying some of those rhetorical challenges and opportunities. By recognizing that documents have genre structures, writers can gain a good sense of the audiences and tasks associated with those documents. Furthermore, by uncovering some of the affordances created by DITA-structured approaches to writing, we can ask students to see what new rhetorical possibilities and challenges await them. Learning to write content with bilocational and translocational meaning poses challenges that make writers look differently at readers and reading, of course, but also at the macrostructures (documentation sets, documents, sections) and microstructures (paragraphs, sentences, phrases) to see how we infuse that content with a particular sense of audience, context, and use. With CMS-generated content, those audiences, contexts, and uses are multiple, and it is worthwhile to gain a suitable vantage point on those changes. Instruction in DITA, in the context of a technical communication course requires teachers and students alike to grapple with the affordances and constraints of the tool.

Even if a class can discover the rhetorical creativity afforded by DITA-structured authoring and required by new reading and use situations, another challenge will be the usability of the tools (see Sapienza, 2004). Notably, Stewart Whittemore (2008) has pointed to the problems with CMS files, DITA included, that make it difficult for writers to rely on their sense of text when writing. The many hundreds of files that might be generated in the process of DITA-structured authoring pose a significant challenge to memory and invention (Whittemore 2008). The difficulty is not debilitating for writing. After students learn the basic lessons outlined, the next set of learning objectives is content management. Students quickly discover the importance of labeling files, storing them in appropriate folders. But they also must learn the importance of giving IDs to files and fragments of content that clearly signal the kind, purpose, and location of information. These are ongoing pedagogical challenges, but ones well worth undertaking.

References

Andersen, R. (2013). Rhetorical work in the age of content management: Implications for the field of technical communication. *Journal of Business and Technical Communication*. doi:1050651913513904.

Andersen, R., & Batova, T. (2015). The current state of component content management: An integrative literature review. *IEEE Transactions on Professional Communication, 58*(3), 247–270. doi:10.1109/TPC.2016.2516619.

Andrade, O.D., & Novick, D.G. (2008). Expressing help at appropriate levels. In *Proceedings of the 26th Annual ACM International Conference on Design of Communication* (pp. 125–130). New York, NY: ACM. doi:10.1145/1456536.1456562.

Bakhtin, M.M. (1981). *The dialogic imagination: Four essays.* (M. Holquist & C. Emerson, Trans.) (Revised ed.). Austin, TX: University of Texas Press.

Batova, T., Andersen, R., Evia, C., Sharp, M.R., & Stewart, J. (2016). Incorporating component content management and content strategy into technical communication curricula. In *Proceedings of the 34th ACM International Conference on the Design of Communication* (Article 35). New York, NY: ACM. doi:10.1145/2987592.2987631.

Bellamy, L., Carey, M., & Schlotfeldt, J. (2011). *DITA best practices: A roadmap for writing, editing, and architecting in DITA.* Upper Saddle River, NJ: IBM Press.

Blythe, S., Lauer, C., & Curran, P.G. (2014). Professional and technical communication in a Web 2.0 world. *Technical Communication Quarterly, 23*(4), 265–287.

Brumberger, E., & Lauer, C. (2015). The evolution of technical communication: An analysis of industry job postings. *Technical Communication, 62*(4), 224–243.

Clark, D. (2008). Content management and the separation of presentation and content. *Technical Communication Quarterly, 17*(1), 35–60. doi:10.1080/10572250701588624.

Doherty, S. (2017). Leveraging industry onboarding materials in the curriculum. *Proceedings of the Annual Conference on Design of Communication SIGDOC 2017*, 1–2.

Evia, C. (2019). *Creating intelligent content with lightweight DITA.* New York, NY: Routledge.

Evia, C., & Priestley, M. (2016). Structured authoring without XML: Evaluating lightweight DITA for technical documentation. *Technical Communication, 63*(1), 23–37. doi:10.1109/TPC.2016.2516639.

Evia, C., Sharp, M.R., & Pérez-Quiñones, M.A. (2015). Teaching structured authoring and DITA through rhetorical and computational thinking. *IEEE Transactions on Professional Communication, 58*(3), 328–343. doi:10.1109/TPC.2016.2516639.

Farkas, D.K. (1999). The logical and rhetorical construction of procedural discourse. *Technical Communication, 46*(1), 42–54.

Hansen, W.J., & Haas, C. (1988). Reading and writing with computers: A framework for explaining differences in performance. *Communications of the ACM, 31*(9), 1080–1089. doi:10.1145/48529.48532.

Hart-Davidson, B. (2005). Shaping texts that transform: Toward a rhetoric of objects, relationships, and views. In C. Lipson & M. Day (Eds.), *Technical communication and the world wide web* (pp. 27–42). Mahwah, NJ: Lawrence Erlbaum.

Kohlhase, A.E., & Kohlhase, M. (2009). Modeling Task Experience in U. A. Systems. In *Proceedings of the 27th ACM International Conference on Design of Communication* (pp. 135–142). New York, NY: ACM. doi:10.1145/1621995.1622021.

Lanier, C.R. (2009). Analysis of the skills called for by technical communication employers in recruitment postings. *Technical Communication, 56*(1), 51–61.

Leijten, M., Van Waes, L., Schriver, K., & Hayes, J.R. (2014). Writing in the workplace: Constructing documents using multiple digital sources. *Journal of Writing Research, 5*(3), 285–337. doi:10.17239/jowr-2014.05.03.3.

Mirel, B. (1998). "Applied constructivism" for user documentation alternatives to conventional task orientation. *Journal of Business and Technical Communication, 12*(1), 7–49.

Morville, P. (2005). *Ambient findability: What we find changes who we become*. Sebastopol, CA: O'Reilly Media.

Priestley, M. (2001). DITA XML: A reuse by reference architecture for technical documentation. In *Proceedings of the 19th Annual International Conference on Computer Documentation* (pp. 152–156). ACM.

Rockley, A., Kostur, P., & Manning, S. (2003). *Managing enterprise content: A unified content strategy*. Reading, MA: New Riders.

Sapienza, F. (2004). Usability, structured content, and single sourcing with XML. *Technical Communication, 51*(3), 399–408.

Schryer, C., Afros, E., Mian, M., Spafford, M., & Lingard, L. (2009). The trial of the expert witness: Negotiating credibility in court documents in child abuse cases. *Written Communication, 26*(3), 215–246.

Swarts, J. (2010). Recycled writing: Assembling actor networks from reusable content. *Journal of Business and Technical Communication, 24*(2), 127.

Van der Meij, H., & Gellevij, M. (2004). The four components of a procedure. *IEEE Transactions on Professional Communication, 47*(1), 5–14. doi:10.1109/TPC.2004.824292.

Van der Meij, H., Karreman, J., & Steehouder, M. (2009). Three decades of research and professional practice on printed software tutorials for novices. *Technical Communication, 56*(3), 265–292.

Whittemore, S. (2008). Metadata and memory: Lessons from the canon of memoria for the design of content management systems. *Technical Communication Quarterly, 17*(1), 88–109. doi:10.1080/10572250701590893.

Appendix: A Visual Summary of DITA Concept, Task, and Reference Topics

Concept

Hierarchy of Concept Topic Elements

<concept>	<title>					
	<shortdesc>					
	<prolog>					
	<conbody>	<p>			<term>,<cite>, <varname>,<sub>,<sup>,<fn>	
		<section>	<title>	<p>		
		<fig>	<title>			
			<image>	<alt>		
		<image>				
		<note>				
		<codeblock>				
		<codeph>				
		<lq>				
		<ul>,<ol>,<sl>				
		<dl>			<dlentry>	<dt>
						<dl>
		<simpletable>	<sthead>	<stentry>	<strow>	<stentry>
				<stentry>+		<stentry>+

FIGURE 9.2 A visual summary of a DITA concept

Task

Hierarchy of Task Topic Elements

<task>	<title>					
	<shortdesc>					
	<prolog>					
	<taskbody>	<prereq>				
		<context>				
		<steps>	<step>	<hazardstatement>	<typeofhazard>	
					<howtoavoid>	
				<cmd>	<menucascade>	<uicontrol>
					<uicontrol>	
				<choices>	<choice>	
				<choicetable>	<chhead>	<choptionhd>
						<chdeschd>
					<chrow>	<choption>
						<chdesc>
				<substeps>	<substep>	<cmd>
				<stepxmp>		<object>
				<stepresult>		
				<info>	<note>	
			<stepsection>			
		<result>				
		<example>				<object>
		<postreq>				

FIGURE 9.3 A visual summary of a DITA task

Reference

Hierarchy of Reference Topic Elements

<table>
<tr><td rowspan="14"><reference></td><td colspan="6"><title></td></tr>
<tr><td colspan="6"><shortdesc></td></tr>
<tr><td colspan="6"><prolog></td></tr>
<tr><td rowspan="11"><refbody></td><td colspan="2"><section></td><td><title></td><td colspan="2"><p>, <table>, <simpletable>, <ul>
<ol>, <dl></td></tr>
<tr><td colspan="5"><example></td></tr>
<tr><td colspan="3" rowspan="4"><refsyn></td><td colspan="2"><fig>, <image></td></tr>
<tr><td colspan="2"><codeblock></td></tr>
<tr><td colspan="2"><synph>, <var></td></tr>
<tr><td colspan="2"><simpletable>, <table></td></tr>
<tr><td rowspan="2"><simpletable></td><td rowspan="2"><sthead></td><td><stentry></td><td rowspan="2"><strow></td><td><stentry></td></tr>
<tr><td><stentry>+</td><td><stentry>+</td></tr>
<tr><td rowspan="3"><properties></td><td rowspan="3"><prophead></td><td><proptypehd></td><td rowspan="3"><property></td><td><proptype>+</td></tr>
<tr><td><propvaluehd></td><td><propvalue>+</td></tr>
<tr><td><propdeschd></td><td><propdesc>+</td></tr>
</table>

FIGURE 9.4 A visual summary of a DITA reference

PART IV
Community

10

EXTENDING THE WORK OF WRITING STEWARDSHIP

Managing Texts, People, and Projects

William Hart-Davidson and Benjamin Lauren

MICHIGAN STATE UNIVERSITY

Chapter Takeaways

- Readers will understand writing stewardship as providing a new value structure for writing expertise.
- Readers will understand content management and project management (PM) as complementary skillsets for writing stewardship.
- Readers can build new learning outcomes and activities into their course designs using the case examples and assignment prompts presented here.

Addie arrived at work to discover she was being promoted. Her new title would be Lead Information Developer, and she would be asked to manage content strategy, teams of people, and new initiatives for her organization. She was taking over for her former manager, who had just retired. The parting advice she received was, "To do this job well, you must be an effective leader, which means you have to be an effective communicator." She mulled over this advice as she moved into her new office and signed on to her computer for the day. Addie took swift action. She decided to call a meeting with everyone on the team to say hello and orient the department, but also to institute some changes to help improve their overall communication. Having spent the last several years of her career at Wild Corp, Addie knew all too well where the bottlenecks in the department's processes were and in what ways the tools they used to coordinate work created as many problems as they solved. It was time to change how the department communicated, and she had a solid sense of where to begin.

Addie decided to make three immediate changes. The first was to centralize communication using a new PM tool. She noted that the department relied heavily on email to share information, which made it easy to overlook time-sensitive project needs. The new tool would make it easier to chat about project work and store information. It would also store the chat information in perpetuity, making it easier to cycle people off and on projects and to make project work more transparent across the team. Second, she decided that it was time for the team to begin practicing Agile Scrum (see the Afterword for more on Agile methodologies). The goal for using Scrum would be to coordinate information about project work and to help the team more quickly solve problems together, rather than in isolation. Her department was made up of about ten people, so she felt a daily Scrum would be a practical way to foster more effective communication. Third, she implemented a new content management system that would allow for people all across the organization to participate in the development of content. This new system would facilitate more efficient reviews and better collaboration, and would also act as a central content library.

What surprised Addie is that the changes she chose to implement, while informed and useful, were not well received by the department during their first meeting. She expected people to be excited for the changes because they would solve so many of their problems. However, the group felt wary of the changes and acted somewhat dismissively and defensively. The timing seemed suspicious to them, and some felt as if they were being forced out of the department. What Addie did not realize when walking in to that meeting is that she inherited both legacy systems of the department, one social and the other technical. She did not realize how important it was to involve the department in her thinking about processes and tools. As the new team lead, her new responsibilities included ensuring that the department, as a whole, communicated consistently well with the rest of the organization.

As the careers of writing professionals advance beyond entry level, they can find themselves in leadership roles in organizations where they embody the role of what Hart-Davidson (2013) calls "writing stewardship," noting that it involves using one's expertise in writing to help others in an organization write well (p. 70). In this chapter, we extend the discussion of writing stewardship to provide a detailed account of the PM knowledge and skills that it necessarily involves. The chapter explains that the work of writing stewardship requires implementing effective PM skills and processes by drawing from one's communication expertise. We will also explain how being a writing steward involves important PM practices, such as negotiating buy-in and facilitating effective teams and teaming.

How (and When) Should you Use this Chapter?

For both new and experienced instructors, it can be a challenge to balance learning goals in a writing course in order to include those that Gagne (1985) calls "attitude" or value changes alongside intellectual skill and knowledge-oriented outcomes. But changing students' values and attitudes about writing can be the most important thing we do. The material we present here—case examples and connections to skills associated with both content management and PM—is meant to help instructors frame learning outcomes like these (taken from Hart-Davidson's syllabus): "Students will understand and value the work to create information assets that are reusable," and

> students will see and appreciate categories of work that align with writing stewards: creating templates, establishing editorial guidelines, creating metadata formats, fostering peer review processes, and other tasks that relate to the management of content and not necessarily to its origination.

These learning outcomes align with acquiring skills and knowledge, to be sure, but are more foundational. We want students to be able to appreciate and represent the work of content and project management as valuable to writing in organizations. So, we ask them to practice those things by putting them in situations where they have to describe, summarize, and defend these practices as valuable. They also need to do them, of course. But just being able to do them is not always enough if we want students prepared for workplaces where their expertise as writing stewards might be needed, but not explicitly understood.

We think the best time to use this chapter, as an instructor, is when you are planning your course. Use it to make sure you attend to attitude and value-oriented learning goals along with skill and knowledge-oriented ones. Use it to plan exercises and perhaps a whole project focused on helping students to understand their knowledge and skills to be useful in the role of a writing steward, helping others within an organization to participate in distributed writing work and to be effective. And, finally, use it to imagine new kinds of deliverables—reflective statements, a formal collaborative project "charter" (see below)—that might fit well into your existing projects and allow you to focus students' attention on writing stewardship.

Why Content Stewardship? Content Management Enables Distributed Work

How does content management and the widespread adoption of content management systems pave the way for writers to become managers of people, texts, and projects? The short answer is that content management systems make it possible for more people to write and for more of the writing that people do to be reused, and that requires organization. This shift—making writing activity distributed in time and space—puts a premium on another category of work: building and maintaining relationships among the people who create, use, and reuse content assets (Clark, 2007; Hart-Davidson, Bernhardt,

McLeod, Rife, & Grabill, 2007; Andersen, 2014; Andersen & Batova, 2015). Just because more people in an organization write, that does not mean that "writers"—those with expertise and/or formal training as professional communicators—are displaced. Rather, their jobs become focused on writing, the verb, as a high-value activity the organization must execute. And this is where PM comes in.

The role of the project manager in organizing groups of writers is to manage both the technical and social elements of work (see Levin, 2010). As we saw with Addie in the example that opened this chapter, the social and technical elements of work are often intertwined in ways difficult to predict. To parse such issues, Stan Dicks (2004) and Connie Kampf (2006) have suggested a rhetorical approach for thinking about the communication of project managers and leaders. In this work, we learn that rhetorical approaches can help those in leadership and management roles strategize about how to communicate effectively. In particular, focusing on the rhetorical situation helps managers analyze and respond to emergent relationships between audience, purpose, and a given message. For example, returning to Addie's dilemma about the audience feeling disoriented by the good-intentioned changes she wanted to make to help manage content more effectively, a rhetorical approach would help to imagine the potential responses to the changes she would want to make. The strategizing and imagining outcomes ahead of time might have saved Addie from making a misstep so early in her role.

Participatory approaches to managing texts, people, and projects are particularly important because they position the audience to be involved in the decisions about work. Finding roots in participatory design, participatory rhetoric extends the idea that communication systems at work contain many of the same social implications as the technologies we develop and contribute to as technical communicators (Dura, Singhal, & Elias, 2013; Lauren, 2018a). In other words, when writers use a particular content management system, that system comes with certain affordances and constraints that influence how people work and share in work. The same can be said about the communication systems project managers use. In this way, Scrum introduces its own logic and rationale that might or might not be helpful or inclusive of all members of the team. As a result, Addie may have unintentionally excluded some members of the team by not involving them in decisions about the social elements of work.

Particularly important to managing content across teams is also how workplaces are organized today. Many organizations that employ technical communicators use distributed teams, where individuals are connected through digital technologies resembling social media. These teams tend to merge or blur the technical and social elements of work out of necessity to move efficiently in tandem in order to remain customer-oriented and competitive in the marketplace (Brafman & Beckstrom, 2006; Rainie & Wellman, 2012; Spinuzzi, 2015). In these contexts, it is useful for project managers to work as facilitators of writing stewardship, which means helping teams to organize themselves in a variety of ways, such as helping people troubleshoot technology issues or giving feedback on writing.

Teaching Essential Strategies of Writing Stewardship

This chapter aims to help instructors teaching content management to prepare students for their new roles as writing stewards. We approach this by drawing upon original research from the authors, and present a set of scenarios that showcase the ways PM knowledge becomes central to the work of content strategy. These scenarios present teaching cases (and accompanying exercises or projects). Each scenario draws on a case study from an empirical study of workplace activity with three features of value to those teaching or preparing to teach technical and professional writers: (1) a framing problem, presented as a short narrative; (2) an analysis of technical and professional writing strategies appropriate for the scenario, situated in the field's literature; and (3) a discussion of the PM knowledge and skills implicated in the scenario, with appropriate links to the PM literature. Our aim in presenting each scenario is to forge links between two scholarly communities that have only occasionally overlapped: technical and professional communication and PM. Additionally, it is important to note that the deliverables of PM work in these situations are not always traditional written genres (e.g., planning documents or budgets). Sometimes, the deliverable is introducing processes that lead to shared involvement in development of such documents and the development of relationships necessary to build trust so that these processes are sustainable and effective.

We chose the three scenarios at the heart of our teaching section because they fall into a category that students, both those preparing for careers as technical and professional communicators and those preparing to teach technical and professional communication, experience as "things we did not know we needed to know." All three are, we argue, essential practices for writing stewards.

Managing Distributed Writing as a High-Value Practice

1. As a writing steward, most of the writing one works with in an organization with a distributed content strategy comes from subject matter experts (SMEs). Their main concern may be its technical accuracy, but its palatability for all the different audiences your organization must deliver to is less obvious for SMEs. This responsibility belongs to the professional writer serving as a writing steward.

Case Example 1

Addie has encouraged MDCorp's information developers to begin thinking about how product documentation could be reconceived as multimedia content—especially short videos—distributed across a customer's experience with the organization. As a result, the writers at MDCorp have been working to gain expertise with their customers' experiences through assembling journey maps, and began to reimagine existing instructional documentation in the form of just-in-time media. This reimagining of content created some complex writing situations for the team. For example, some of the developers felt it was far easier to take SME writing or interview data and help to make it accessible to the audience in text form than to depict similar information in a video or illustration. Others had questions about how instructional information, some of which would likely change with the rollout of new software updates, would quickly make multimedia content out of date and far harder to update. Discovering a content strategy that could respond to these challenges and finding a repeatable process that could be managed became a central concern for the team—one that inspired a fair amount of debate.

To help facilitate these changes, the department began working with a design thinking methodology to respond to pain points in the customer experience. They also began to collaboratively map out the customer's journey with products and services offered by MDCorp. Meanwhile, some developers began to experiment with multimedia texts, such as videos. As they did this work, they also implemented the new PM system Addie had introduced. This new system could be used to replace email while also providing space for people to share and retrieve information. Because the team often worked with people distributed across offices in different time zones, the PM system could help to efficiently coordinate the development of content.

Technical and Professional Writing Strategies in Focus

In the book *Writing Workplace Cultures: An Archaeology of Professional Writing*, Jim Henry (2000) draws on observations of more than eight writers in a wide variety of workplaces in order to understand not *what* but *how* they write every day. Henry reports something quite astonishing to many: the work of professional writers includes a lot of activities that are not recognizable as "writing," such as interviewing SMEs and learning new kinds of media production techniques to be able to produce audio, video, infographics, or websites. And lots of research and inquiry, from customer surveys to competitive analysis of products or services. The amount and types of these kinds of activities, according to Henry, varied quite widely too, meaning that there was no one clear way to be (or even to prepare to be) a "professional writer." But there was one activity Henry saw in common across all the case studies he analyzed: writers must be continually learning and typically have a key role to play in their co-workers' learning.

Inspired by studies like Henry's (2000), other researchers have sought to add detail to the portrait of technical communicators' work lives as it relates to learning and what affords the ability of writers to be successful in the face of a near-constant demand to update skills, become better researchers and interviewers, learn new software, and grow capacity in different types of media. Jason Swarts' (2017) study of workplace writers in *Together with Technology: Writing Review, Enculturation, & Technological Mediation* offers detailed portraits of early-career professionals learning from their more experienced peers. A key takeaway from Swarts' study is the value of feedback—giving and receiving—as perhaps the signature component of a professional writer's career success. Early on, the ability to seek out and learn from feedback provided by others—including more experienced communicators as well as SMEs, supervisors, clients, and customers—is the skill that promotes individual development and accelerates career advancement. Later, it is the ability to provide good feedback and ensure that others are using feedback to make improvements to their work that distinguishes professional writers as emerging leaders who can take on more responsibility.

Project Management Knowledge in Focus

This sort of coordinative work across distributed and cross-functional teams requires a type of work we call "staging" or "planning" in PM. During staging, a project manager works with the team to create a game plan for making sure work gets done effectively and on time. The project manager's role in this situation is to learn how to best design a workflow that makes room for people on the team and corresponding stakeholders to participate in the work that needs to be done. During staging, the project manager also coaches the SMEs to use their expertise to help determine what needs to be done while also working to get "buy-in" from the team. To negotiate buy-in, the project manager must communicate to the team that everyone must find a way to work together for the project and for the group to be successful. To work together, the team must also agree on a workflow and stick to it throughout the duration of the project.

As we can see with the team at MDCorp, the PM system was being used to centralize a workflow for everyone on the team. The workflow was used to coordinate and cultivate productive methods of exchanging information, but in cooperative ways. Many systems are used to centralize workflow in this way. For example, teams can use content management systems alongside ticketing systems that help establish task lists and implement a content strategy in an organization. Important to this process is that PM is only successful if teams and SMEs buy-in to the process. In other words, teams must learn to effectively participate in project work to be successful.

2. The writing one does as a writing steward looks different from the writing aimed at external audiences. Put simply, the writing steward makes texts that help others communicate. We call these "coordinative genres," following Spinuzzi, Hart-Davidson, and Zachry (2006).

Case Example 2

Writing stewards lead through and by communication actions. Importantly, their leadership approach also influences how they embody the role of a writing steward. Metaphorically, we can understand the influence of leadership approach on writing stewardship as inhabiting the role of a *gardener* or a *chef* (see Lauren, 2018a). As a leader, *gardeners* tend to use coordinative genres to model and cultivate the kind of communicative behaviors that would be useful to their team. For example, a gardener might use coordinative genres to get feedback on the implementation of a new calendar system for the team. To work toward consensus for the new calendar system, a gardener might take communication actions such as moderating group discussions, developing feedback surveys, or requesting other forms of informal feedback. The goal for approaching the work in this way is to lead by cultivating conditions for success.

Chefs, on the other hand, often use coordinative genres as a means for implementing recipes for achieving a project's success. Chefs lean on formal training to choose or develop processes, tools, and procedures that can help a project team be successful. For example, a chef might rely on established process improvement procedures to help teams become more successful. Their focus might be on addressing alignment issues up, down, and across an organization's hierarchy as a way to seek clarification about a project's goals. They do this work through writing mundane, coordinative genres to establish and clarify ideas and make sure they are represented in key terms and phrases on project documentation, such as charters.

Different workplace environments require a leadership performance by the writing steward that can be understood as acting as either a gardener or a chef. Both approaches use coordinative genres to help lead teams, but they also use them to shape the team's view of and approach to doing project work. This kind of writing stewardship may seem invisible to the external audiences of an organization, but it directly impacts how people work and how content is developed and shared. In Addie's case, she understood her role was to approach the people management elements of her job as a gardener and the technical side as a chef.

Technical and Professional Writing Strategies in Focus

A curious dynamic in organizations is a condition when written communication is important to the overall work and mission and people with the title of Writer or those who have expertise or degrees in "technical writing" might find themselves doing very little writing! Or, at least, very little of what many people think about as writing: sitting in front of a screen, drafting text. Instead, these professionals ensure that the whole organization is writing and doing it well (Conklin, 2007).

It then falls to those with expertise in writing to help create, train, and stabilize effective writing processes and workflows, making the work of producing high-quality writing visible and making the products of that work usable by the extended network of content providers and reviewers who engage in it. So, writing stewards might not draft the text that goes into an engineering report, but they might create the templates people use to scaffold processes and ensure consistency. They establish the criteria for and coordinate various stages of review—by SMEs to ensure quality, by accessibility experts to ensure the work will reach its intended audiences, etc. Writing stewards also do the hard work of ensuring that the text and visual assets an organization creates can be adapted and reused across multiple media formats—in print, online, via mobile—across any number of necessary genres and audiences.

Johnson-Eilola's (1996) influential article "Relocating the Value of Work: Technical Communication in a Post-Industrial Age" predicted the shift toward writing stewardship that we now see as the status quo in many organizations as technical communicators' knowledge and expertise moved closer to the core value of what the economy values in a post-industrial era. The argument Johnson-Eilola made just as the world-wide web was taking shape felt, to many, counterintuitive: the more people who do not have the title of Writer need to write as part of their work, the more valuable "writers" will become in organizations.

The corresponding shift, as Johnson-Eilola's (1996) title makes clear, is that those with specialized expertise in writing do less product-focused or what economist and former Secretary of Labor Robert Reich (1991) termed "routine production" work, and more "symbolic-analytic" work. Symbolic-analytic work asks writers to work in abstractions—methods, processes, organizational strategies, quality-control criteria and standards—exercising their knowledge to deliver information in an increasingly service-oriented vs product-oriented economy. Why does this kind of work have more value? Because it ultimately allows production work to be more broadly distributed in time and space, enabling access to more customers, members, and students and providing a means for previously local businesses to achieve global scale. But distributed work requires enhanced coordination, and someone whose knowledge and experience can attend specifically to keeping distributed teams communicating, together, effectively.

Project Management Knowledge in Focus

As practices become more distributed across teams and organizations, coordinating communication becomes far more complex. New kinds of genres become important: journey maps, swim-lane diagrams, and project charters are examples. They can not only help prepare a team, but they also permit the writing steward to stay abreast of the team's progress, highlighting where they might need to intervene to cultivate better conditions or follow a better recipe for success. In other words, the writing steward leads and communicates effectively by taking action and responding to the exigence of a situation and context. Sometimes that means providing templates or practices that others can adopt into their workflow. In short, writing stewards design communication patterns for others to follow.

Coordinating communication is nearly constant throughout the development of any team-based project, and doing this work well requires situationally-responsive approaches. Writing stewards have to know what kind of leadership strategy they will deploy as communicators, and how the approach they choose fits their workplace context. One way to do this work is by identifying their communication approach as either a gardener or as a chef. Such approaches can help to design communicative patterns and values that help teams ultimately become successful.

3. A writing steward often finds that they have a new priority that is more important, even, than shipping high-quality communication products: keeping a content team healthy, happy, and productive. A team that gets a great product out the door once and destroys itself in the process has not failed. But the work required of the writing steward to keep teams together is often invisible.

Case Example 3

Change is constant in our workplaces today. Organizations rethink their structure, teams change, new businesses are acquired, and people shift in and out of roles. Even the technology many people use to support our work today seems to constantly be changing via updated hardware and software. With all this change circling around project teams, it becomes imperative that teams learn how to build and maintain relationships for the sake of sustaining and maintaining their short- and long-term success. Addie saw this situation was the case at MDCorp—that she decided to make changes without building relationships in her new role as the organization's lead. Perhaps this decision was one reason the department resisted her first attempts at implementing change. Indeed, restructuring can cause quite a panic and even resistance (Lauren, 2018b; see also Suchan, 2006).

At MDCorp, a team of information developers worked through the changes Addie introduced after they spent some time sorting through why the changes were needed and how they would be implemented. At first, Addie worried that morale among employees was somewhat low, and relationships suffered. Morale was low because some felt the change process was unclear, which degraded trust within different circles of employees. She worried that her approach to implementing the changes made people worry they were not part of the future at the organization. After spending more time discussing the issue, the people on her team began to see how the changes benefited their project work more clearly. But it took time and a great deal of carefully constructed communication. Addie's work turned toward building stronger relationships and trust among her team members. Rather than a distraction from producing a better product, she understood that it was the path to helping the team reach that goal.

Technical and Professional Writing Strategies in Focus

Discussions of ethics and ethical practice in technical and professional writing have often been limited to the ways writers elect to represent others (e.g., Katz, 1992) or the ways they represent knowledge and/or the truth of a situation given the available evidence (e.g., Herndl, Fennell, & Miller, 1991). Implicit in these discussions, however, is the role of the writer as an organizational *agent* with the ability and responsibility to act in ways that preserve human relationships and, occasionally, human lives even if these actions mean challenging organizational practice or structures. A writing steward learns to employ *phronesis*—the art of ethical decision making—to create the kind of workplace relationships that enable both high-quality work and healthy interactions among co-workers.

In her article "A Humanistic Rationale for Technical Writing," Carolyn Miller (1979) provides the foundational case for a view of writing as inclusive of ethical decision making. Miller's case for technical communication as a humanistic practice is grounded in an orientation toward scientific and technical knowledge that breaks from the positivist view that truth is discovered rather than constructed by way of scientific practice, including communication. In the positivist view, communication is merely transmission of what is already known, preferably in as clear and concise a manner as possible. Miller argues that a more rhetorical view of science recognizes the role of persuasion and assent in the way scientific facts are proposed, reviewed, and either accepted or rejected by various expert and lay audiences. In this framework, how to present information to best reach and win the assent of these various audiences requires rhetorical knowledge and the kind of practical wisdom to make both effective and ethical choices.

Miller's (1979) argument is pitched to her colleagues in English Studies skeptical that a course in technical communication can "count" as a Humanities elective. She contends that it can and should, based on the way it positions the writer to develop and draw upon the sort of practical wisdom to make ethically responsible choices. All too often, though, this component of a technical communicator's work and expertise is subsumed by what Scott (2004) calls the "hyperpragmatism" of both learning institutions and workplaces who frame the development of the skills of technical writers in an instrumental fashion—as things that build value in and for companies and organizations. Scott, like Miller, sought to create an alternative rationale for technical communication learning and, in particular, for a common type of learning experience in technical communication programs—service-learning projects—beyond the skills they engender. Scott's argument aligns with ours, suggesting that what can be most valuable for students who engage in service learning is the ability to nurture relationships, build trust, and engage in participatory decision making with community partners. These experiences and the qualities they help students develop exceed the instrumentalist logic of hyperpragmatic logic, approaching the humanistic ends that Miller wrote about 25 years earlier.

Rebecca Pope-Ruark (2014) provides, for us, the direct connection to the idea of managing people as an important expression of technical communicators' knowledge. Building on the foundation provided by each Miller and Scott, Pope-Ruark argues for three higher-order forms of knowledge that we readily associate with people, projects, and texts. These terms are "*phronesis*," "praxis," and "*metis*." For Pope-Ruark, the three are interconnected and speak to the ability of a communicator to make well-reasoned and responsive choices (*phronesis*), mobilizing their knowledge and experience in a practical way (praxis) and doing so strategically, matching techniques with the needs of a particular project, audience, and situation. Using Agile or Scrum-based PM, Pope-Ruark argues, is an ideal way to foster *metic* expertise in conjunction with *phronesis* and praxis. A key reason is that Agile/Scrum methods explicitly foreground building relationships among team members that build trust and allow for the team to learn and grow together, building the team's overall capacity by encouraging the development of the individuals on it. Indeed, these strong relationships are what enable a team to become more agile.

Project Management Knowledge in Focus

Although it might be new for technical and professional writers to think of relationship-building as a priority, PMs are used to thinking about the integrity of the working relationships in an organization as something they can facilitate. PMs have strategies for conducting meetings, managing workloads, and communicating in both formal and informal ways to keep working relationships intact even as the team keep their own focus on the quality of their work product. Many of these strategies are rooted in keeping people involved, particularly those stakeholders who are not necessarily doing the work of the project. Making sure to pay attention to the health of relationships on teams is essential to a project's success, particularly during times of change.

Additionally, to help facilitate relationship-building, PMs must pay attention to the power dynamics present in different organizational and team structures. These power structures might make it difficult for some people to feel "safe" to participate in different interactions and communication situations. To address these power dynamics, PMs can help facilitate team interactions using activities such as creating rules for interactions during meetings, holding informal meetings, or practicing, and teaching others to practice, active listening techniques.

Activities and a Vision for a Career-oriented Vision of Writing Stewardship

We close the chapter with a set of exercise prompts that teachers of technical and professional writing can build into their assignments that help to emphasize the essential practices of writing stewardship. Each activity includes a focus on PM knowledge and genres along with a reflective component that asks students to examine how they understand their success criteria to include people and projects as well as texts. Depending on the course, you can do these activities in a single session or scale them up to become full projects with deliverables.

- Ask students to select a workplace genre with which they are familiar as a writer and prepare a brief training session that would allow a colleague who plays a different role in the organization to learn to create or contribute to writing a new instance of that genre. For example, the student might create a tutorial about how to create "release notes," a genre that introduces customers to new features of a product being rolled out. The goal of this exercise is to capture not only the textual features of the genre but also a full sense of the process—how does the writer learn about the new features *before* release (e.g., an SME interview? Directly reviewing a pre-release version?), how does she choose the right media formats to present information for the target audience (e.g., do we need a video? Static or dynamic screen captures?), who reviews the draft and suggests revisions, how are drafts approved for release and publication, how can assets produced in the drafting of this particular instance be reused as part of a broader content strategy (e.g., in social media promotions of new features?).
- Ask students working on a group project to begin by drafting a project charter, a document that asks them to think explicitly about the project's goals in relation to the individual team members' learning goals, roles, and responsibilities on the team. Students might further craft a charter that incorporates the four elements of the "Agile Manifesto" that Rebecca Pope-Ruark (2014) writes about as an appropriate ideology for developing metric intelligence in technical communication by foregrounding intra-team expectations and

relationships: (1) individuals and interactions over processes and tools, (2) working software over comprehensive documentation, (3) customer collaboration over contract negotiation, (4) responding to change over following a plan.

- Ask students to facilitate a disagreement between two employees but do so using a technology such as video chat or instant message. The goal here is to experience first-hand how blurring technical and social elements influences the rhetorical communication approaches of those who manage writers. Students will see how the environments we use to communicate across distance influence the way we shape messages (Swarts & Kim, 2009), as well as how digital technologies continue to shift how and when we work (Rainie & Wellman, 2012). To help prepare students to do this exercise, you should write out a scenario and prepare students to moderate the session in a goal-oriented way (see Lauren, 2018b, for an earlier iteration of this exercise).

Conclusion

Preparing students to be writing stewards looks beyond the entry-level position responsibilities that often become the focus of our curricula to a set of practices that help young professionals develop fulfilling careers and work lives. This career-oriented approach (Jablonski, 2005) also serves students well as they prepare for a world with rapidly changing work environments, new job categories, and new technologies. Emphasizing rhetorical approaches to thinking about the social and technical elements of managing texts, people, and projects helps us understand how project managers and content strategists can extend their work as writing stewards, and act as leaders by helping everyone in their organization contribute to content creation in unique ways. When extending this work in a PM role, it is important to emphasize that the deliverables of this work have become inclusive of many other kinds of emerging workplace genres, such as introducing new processes for managing work, building and maintaining relationships, and learning how to facilitate the social elements of communicating in the workplace.

References

Andersen, R. (2014). Rhetorical work in the age of content management: Implications for the field of technical communication. *Journal of Business and Technical Communication, 28*(2), 115–157. doi:10.1177/1050651913513904.

Andersen, R., & Batova, T. (2015). The current state of component content management: An integrative literature review. *IEEE Transactions on Professional Communication, 58*(3), 247–270. doi:10.1109/TPC.2016.2516619.

Brafman, O., & Beckstrom, R.A. (2006). *The starfish and the spider: The unstoppable power of leaderless organizations*. New York, NY: Penguin.

Clark, D. (2007). Content management and the separation of presentation and content. *Technical Communication Quarterly*, *17*(1), 35–60. doi:10.1080/10572250701588624.

Conklin, J. (2007). From the structure of text to the dynamic of teams: The changing nature of technical communication practice. *Technical Communication*, *54*(2), 210–231.

Dicks, S. (2004). *Management principles and practices for technical communicators*. New York, NY: Longman.

Dura, L., Singhal, A., & Elias, E. (2013). Minga Perú's strategy for social change in the Peruvian Amazon: A rhetorical model for participatory, intercultural practice to advance human rights. *Rhetoric, Professional Communication and Globalization*, *4*(1), 33–53.

Gagne, R.M. (1985). *The conditions of learning and theory of instruction* (4th ed.). New York, NY: CBS College Publishing.

Hart-Davidson, W., Bernhardt, G., McLeod, M., Rife, M., & Grabill, J.T. (2007). Coming to content management: Inventing infrastructure for organizational knowledge work. *Technical Communication Quarterly*, *17*(1), 10–34. doi:10.1080/10572250701588608.

Hart-Davidson, W. (2013). What are the work patterns of technical communication. *Solving problems in technical communication*. (pp. 50–74). Chicago and London: The University of Chicago Press.

Henry, J. (2000). *Writing workplace cultures: An archaeology of professional writing*. Carbondale, IL: Southern Illinois University Press.

Herndl, C.G., Fennell, B.A., & Miller, C.R. (1991). Understanding failures in organizational discourse: The accident at three-mile island and the shuttle challenger disaster. In C. Bazerman & J. Paradis (Eds.), *Textual dynamics and the professions: Historical and contemporary studies of writing in professional communities*. (pp. 279–305). Madison, WI: University of Wisconsin Press.

Jablonski, J. (2005). Seeing technical communication from a career perspective: The implications of career theory for technical communication theory, practice, and curriculum design. *Journal of business and technical communication*, *19*(1), 5–41.

Johnson-Eilola, J. (1996). Relocating the value of work: Technical communication in a post-industrial age. *Technical Communication Quarterly*, *5*(3), 245–270.

Kampf, C.E. (2006). The future of project management in technical communication: Incorporating a communications approach. Paper presented at International Professional Communication Conference. New York, NY: IEEE Xplore. doi:10.1109/IPCC.2006.320372.

Katz, S.B. (1992). The ethic of expediency: Classical rhetoric, technology, and the holocaust. *College English*, *54*(3), 255–275. doi:10.2307/378062.

Lauren, B. (2018a). *Communicating project management: A participatory rhetoric for development teams*. New York, NY: Routledge.

Lauren, B. (2018b). Preparing communication design students as facilitators: A primer for rethinking coursework in project management. *Communication Design Quarterly*, *6*(3), 59–65.

Levin, G. (2010). *Interpersonal skills for portfolio, program, and project managers*. Vienna, VA: Management Concepts.

Miller, C.R. (1979). A humanistic rationale for technical writing. *College English*, *40*(6), 610–617. doi:10.2307/375964.

Pope-Ruark, R. (2014). A case for metric intelligence in technical and professional communication programs. *Technical Communication Quarterly, 23*(4), 323–340. doi:10.1080/10572252.2014.942469.

Rainie, L., & Wellman, B. (2012). *Networked.* Cambridge, MA: MIT Press.

Reich, R.B. (2010). *The work of nations: Preparing ourselves for 21st century capitalism.* Cambridge, MA: Vintage.

Reich, R. (1991). *The work of nations: A blueprint for the future.* New York, NY: Vintage.

Scott, J.B. (2004). Rearticulating civic engagement through cultural studies and service-learning. *Technical Communication Quarterly, 13*(3), 289–306.

Spinuzzi, C. (2015). *All edge: Inside the new workplace network.* Chicago, IL: University of Chicago Press.

Spinuzzi, C., Hart-Davidson, W., & Zachry, M. (2006). Chains and ecologies: Methodological notes toward a communicative-mediational model of technologically mediated writing. Paper presented at the SIGDOC, 43–50. doi:10.1145/1166324.1166336.

Suchan, J. (2006). Changing organizational communication practices and norms: A framework. *Journal of Business and Technical Communication, 20*(1), 5–47. doi:10.1177/1050651905281038.

Swarts, J., & Kim, L. (2009). Guest editors' introduction: New technological spaces. *Technical Communication Quarterly, 18*(3), 211–223.

Swarts, J. (2017). *Together with technology: writing review, enculturation, and technological mediation.* Abingdon: Routledge.

11

TEACHING CONTENT MANAGEMENT FOR GLOBAL AND CROSS-CULTURAL CONTEXTS

Kirk St.Amant

LOUISIANA TECH UNIVERSITY AND UNIVERSITY OF LIMERICK

Chapter Takeaways

After reading this chapter, instructors should be able to

- Understand how cultural factors affect expectations of credible content.
- Apply globalized rhetoric to identify the credible content expectations of different cultures.
- Use globalized rhetoric to create messages that meet the credibility expectations of other cultures.
- Teach students to use the globalized rhetoric approach to understand, research, and develop materials that meet the credible content expectations of different cultures.
- The objective is to provide readers with a framework they can use to better understand—and teach others about—credibility dynamics affecting how different cultural audiences perceive the credibility of content.

The interconnected nature of modern society means almost all organizations engage in some form of international interaction. Additionally, the rate at which technologies and markets evolve means content created for local audiences will likely move into global contexts at some point (Collins, 2010). At the same time, immigration and demographic trends have made local settings far more multicultural and multilingual than ever before (Milanovic, 2013; Romero, 2017). As a result, cross-cultural communication is increasingly part of one's everyday life. These factors have important implications for individuals working in content management (CM).

Not all cultures use content in the same way, nor do they have common expectations of what constitutes effective or needed content. These factors can create challenges for CM practices on a variety of levels. Addressing such situations involves raising awareness of these factors and understanding how culture affects content consumption. Technical communication instructors can use materials and individuals from other cultures to help students develop the awareness needed to navigate international and cross-cultural situations more effectively. This chapter provides an overview of how cultural aspects can affect perceptions and uses of content. It also provides instructors with sample approaches and activities they can use to introduce students to these ideas.

Content Management and Cultural Contexts

CM is about re-use. One creates content (e.g., text, images, or other forms of information) so it can be re-used in other materials at different times (TCBOK, n.d.). The goal involves minimizing the time spent creating new content by tracking the content already available on a topic. When organizations need to share information, they use the content they have to create different products (e.g., brochures, manuals, websites, etc.) as required. This re-use allows organizations to create new materials quickly and consistently as all content comes from one source (a single source) (Hannon Hill Corporation, 2010).

Although this process seems easy, it can be challenging. Content is not neutral. When shared with others (i.e., published), content needs to be adapted to the expectations of an audience, and these expectations are not always uniform (Barker, 2016). For content to be effective, individuals need to access it readily, interact with it seamlessly, and use it easily to achieve an objective—a process known as "consuming" content (Barker, 2016). This focus on consuming connects content to usability: individuals have to use content—or see it as useful—for it to have value. Otherwise, the time and effort spent creating, editing, and publishing content is for naught. As more entities now produce content and compete for consumer attention, readily consumable content can be essential to an organization's success. Certain factors, however, can affect how audiences from different cultures perceive and use/consume content. If not recognized and addressed, such aspects can adversely affect an organization's success in global markets.

Culture is multifaceted, and it can affect every aspect of human life. It can also have pronounced effects on content expectations, encompassing everything from the languages individuals use when communicating to how persons interact with technologies to exchange ideas (St.Amant & Rice, 2015). These differences can affect CM practices in international (i.e., across nations) and intercultural (i.e., across cultures) settings. Successful CM in such contexts involves understanding cultural dynamics affecting

- How content is created, edited, and published.
- How content is perceived and consumed.

Addressing these areas in CM involves an understanding of rhetoric, or the expectations cultures associate with communication and content.

Rhetoric, Genres, and Content Expectations

Rhetoric focuses on how individuals use content to achieve an objective. Such use can encompass everything from the topics included in certain content to the overall structure or content in different published materials (Berkenkotter & Huckin, 1995; St.Amant, 2006). Central to this process is the audience, and an audience's expectations of genre—the established formats for sharing information (Berkenkotter & Huckin, 1995).

From a CM perspective, audiences use a particular kind of published content—or genre—to achieve a specific objective. (This use is the reason they are consuming that content.) They expect to encounter certain information in that genre; that is, information associated with the purpose for which they are consuming that content (Halvorson, 2008; Barker, 2016). When audiences access content in a manual on how to change a tire (a genre), they expect to encounter information that addresses that topic. If the published content in this genre meets this expectation, it is used/consumed effectively and the audience is satisfied. If not, the audience will likely become frustrated and look for other sources (and perhaps other products) with published content that better addresses their needs.

Information, however, is not the sole element associated with meeting audience expectations of content in genres. So too is structure; that is, how content is organized within a genre/published work (Miller & Selzer, 1985; Berkenkotter & Huckin, 1995). Most audiences expect the content in a genre to be organized in a particular way, and that structure maps closely onto the reasons for which they consume content (Campbell, 1998; Tebeaux, 1999). In a manual on changing a tire, the step order affects how it is performed. If the published content gives audiences the sense information is missing or out of order, they might become dissatisfied with it (Driskill, 1996; Weiss, 1998; Walzer, 2000). If so, they might search for other sources of published content to meet their genre-related expectations.

Genre and Content Management Practices

These aspects of genre are often central to CM guidelines. First, before content can be published, one needs to determine what content an item should contain to meet audience expectations (Hannon Hill Corporation, 2010; Barker, 2016). Individuals must then define the objective an audience associates with the consumption/use of a particular kind of published content (i.e., what the genre for the final published content will be). Next, content managers have to determine the purpose the audience associates with consuming that content and identify what kind(s) of content to include in the final, published entry. Once established, content managers would use these parameters to determine

- What required (based on genre-expectations) content already exists and needs to be organized and edited.
- What new content needs to be created.

Content managers must then organize the resulting content—new and old—into the order audiences associate with the related genre (Scime, 2009; Barker, 2016). The resulting product is then reviewed and edited to conform to audience expectations for the related genre. Through this process, content managers craft content that meets the consumption expectations an audience associates with a genre of published content (Scime, 2009; Barker, 2016).

Genre and Content in Cross-cultural Situations

Theoretically, this process seems relatively straightforward. When moved to international and intercultural contexts, however, a variety of factors need to be considered. For example, not all cultures have or use the same genres to convey information/share content (Woolever, 2001). Rather, genres that seem relatively common in some cultures (e.g., a request for proposal, or RFP), might be of little use, or seemingly non-existent, in another. Without knowing this factor, organizations could inadvertently waste time publishing content a given national or cultural audience will not use.

In other cases, a given genre for published content might exist in another culture, but the members of that culture associate different purpose with that genre (Campbell, 1998; Grundy, 1998; Tebeaux, 1999). Failure to understand such differences could result in published content that fails to provide information an audience expects or needs. Such lack of understanding could also lead to published content that members of a cultural audience see as unnecessary or distracting, because it is not the content they expect to encounter in the related gene. Some cultures, for example, expect marketing content to contain in-depth technical specifications on a product so they can assess the quality of that product; others do not (Ulijn & Strother, 1995). Other cultures might expect instructional content (e.g., an instruction manual) to contain more text, more images, or more text on the specifications of a product. Published content that does not meet these expectations could lead members of certain cultures to view the content—and the quality of the related product—with skepticism (Ulijn & Strother, 1995).

In yet other instances, cultures can have different expectations of the order in which content should appear in a genre (Driskill, 1996). In some cases, this involves how implicitly or explicitly individuals convey ideas in different sections of published content (Driskill, 1996; Campbell, 1998). In others, it can involve when certain information should appear in a genre (e.g., at the beginning or at the end of the related published content) (Campbell, 1998; Grundy, 1998). Failure to meet audience expectations can affect uses of content and perceptions of related products.

Aspects of visual vs verbal and the medium for sharing published content can also vary from culture to culture. So too can the amount of visual vs textual content to include in a genre (Cohn, Taylor-Weiner, & Grossman, 2012). This factor means certain cultural groups might think published items need more visual content to be effective. Conversely, other groups might find a certain level of visual content to be overwhelming. Such factors affect how members of a culture perceive the effectiveness of and are able to use published content. Additionally, what constitutes credible or acceptable visual content can vary from culture to culture (St.Amant, 2005). Such variations can result in unexpected reactions as a result of confusion and offence.

Finally, cultures can have different expectations of how individuals use media to consume content. Cultures, for example, can have different perspectives on what content is appropriate to publish via different social media genres, as well as what content individuals expect to see published via these media (St.Amant, 2015). Additionally, media commonly used to distribute content in certain cultures (e.g., Twitter), might be rarely used in others. And in some cases, certain venues for publishing content might be restricted—or illegal—in a nation or region (St.Amant, 2015). All these factors affect if and how content should be created, edited, and published for different cultural and national audiences.

What makes these various cultural-rhetorical factors particularly challenging is that they are connected to the culture of the audience, not necessarily the language in which individuals consume content. As such, translating content into other languages is not an inherent guarantee that content will meet the rhetorical expectations of different cultural audiences (Esselink, 2000; Yunker, 2003). Likewise, translation does not inherently address expectations cultures associate with visuals or with the uses of different media to access and share published content (Esselink, 2000; Yunker, 2003). These factors exist at a much deeper level and often transcend language. As such, individuals from other cultures use the rhetorical expectations of their native culture to assess content, even when it is published in other languages (Ulijn, 1996). Failure to understand such factors can affect how a national or cultural audience consumes content. It can also affect the international/intercultural successes of organizations whose content fails to address such expectations.

These culture-related factors do not just affect content consumption. They also affect how content is created, edited, and published. This is because many CM practices are distributed in nature. No one individual creates, edits, or publishes all content for an organization. Rather, a central aspect of effective CM is that multiple individuals can contribute, edit, and even publish content at different points in time (Barker, 2016; Optimizely, 2019). Coordinating such distributed practices is challenging when they involve members of a common culture; when expanded to content creators, editors, and publishers from different cultures, variations in rhetorical expectations make these practices more complex.

Guidelines and Structures

To address distributed content-related practices and differing audience expectations, organizations often create guidelines for who may create, edit, or publish content and how they may do so (Maiella, 2015; Rockley, 2016). Such guidelines can include everything from how content should be structured to how it should be compiled and edited to create different genres for publication. These guidelines can also help manage distributed content-creation practices by providing a common structure for how to craft and edit content. Doing so involves adding an element of standardization and predictability to a complex rhetorical situation.

For this approach to work, organizations should create three kind of guidelines. The first involves who can create and contribute new content, and how that content must be formatted (Maiella, 2015; Rockley, 2016). These guidelines help

- Avoid creating redundant content due to multiple individuals providing the same information.
- Facilitate the creation of content that is easy to edit for future audiences and formats.

The objective is to create one source—a single source—of content on a given topic or area. Individuals would then access and use or adapt the content of this single source later as needed. This approach can help in global settings by having only individuals from or familiar with a cultural audience create content for that group. Theoretically, these individuals would know the genre-related factors to address and create content that meets the rhetorical expectations of the related cultural audience.

The next kind of guidelines involves editing existing content (Scime, 2009; Skyword, 2015). Although single-sourced content might be effective as is, that situation is not always the case. It might need to be edited, updated, or adapted to meet the expectations of certain audiences or for distribution across different media (Scime, 2009; Hannon Hill Corporation, 2010; Barker, 2016). Such editing might involve multiple individuals revising different pieces of content for a common publication project. As with authoring, each editor could apply different practices and approaches that might make the resulting entry disjointed or confusing.

Moreover, if individuals do not keep track of edits/revisions over time, it can be difficult to determine if the existing content an organization has is current, has been revised for certain projects, or needs updating. For these reasons, organizations need to develop guidelines for how to edit content for projects (Scime, 2009; Skyword, 2015). Such guidelines create consistent editing practices for individuals collaboratively editing a project. They also provide guidance in tracking such edits so later users could determine the status of the content they accessed (e.g., when was it last updated, and how).

From an international and intercultural perspective, this editing practice can help address rhetorical differences relating to genre and content. Using editors from the same culture as the audience can increase the chances these editors address genre-related cultural differences. These editors can also provide notes on how and why they edited the related content. (Such notes can help future editors know to first review that content carefully as it has been adapted for a particular cultural audience.)

A third set of guidelines would focus on the publishing of content (Scime, 2009; Lieb, 2018). As content creation and content editing can be distributed across individuals, so can the ability to publish or share final content with an audience. Providing multiple persons with the authority to publish final content can speed its release. Done in an un-structured way, however, such publication practices can lead to problems, from organizations not knowing when content is published to having multiple persons publishing different "final" versions of content (Diorio, 2001; Lieb, 2018). To curtail such situations, organizations should adopt guidelines for when content is finalized and ready to publish, who may publish that final content, and how those who publish content make others aware of such publication.

Organizations can use these publishing guidelines to address international and intercultural situations in two ways. First, by restricting final publishing authority to members of the same culture as the consumer, organizations can create a final cultural review to make sure content meets cultural expectations prior to publication. Publishers from the same culture as the audience would review the final content to determine if it addresses the genre-related expectations of that audience. Based on this review, final edits could be done as needed to make sure the published content met cultural-rhetorical expectations.

In a different scenario, the authorized content publisher would consider the nature of the overall project before any content-related work is done. In this case, before an organization undertakes a project, individuals (assigned content publishers) from the same culture as the audience would review it. This review would determine if the genre the organization wished to use exists in and is used by the related cultural audience. If so, the project could move forward; if not, the project would be abandoned before time and energy were dedicated to creating unnecessary materials. This approach would apply to both the kinds of genres created and the medium used to convey content (i.e., determining if members of a culture actually consume content via certain media).

The scenarios described here represent an idealized approach to CM in global contexts. The complexities of coordinating such processes can make them too cumbersome or complex for an organization to address via CM practices alone. For this reason, the task of addressing international and international aspects of CM often falls under the area of *localization*: the process of revising or creating materials to meet the expectations of a cultural group (Batova & Clark, 2015). For such processes, an organization works with a localization company or firm—one that specializes in such cultural custom-fitting—to develop content for a given cultural audience prior to publishing it. This situation is not always the case, and that can cause problems.

Explaining these concepts to technical communication students can seem fairly straightforward. Helping them understand such factors and appreciate why and how to address them can be more challenging. Certain approaches and activities can help provide students with a foundational familiarity with these concepts. They can also provide students with basic strategies for addressing such situations and doing CM work in globally distributed and culturally diverse contexts.

Teaching Focus

Teaching students about CM in international and cross-cultural contexts can be daunting. The key is to make them aware of how cultural factors can affect CM processes. With this knowledge, students can learn to ask more informed questions about the cultural audiences with whom they create and for whom they publish content. They can also begin thinking more reflectively about the globally distributed collaborators with whom they might work on CM projects. Developing this kind of introspection and understanding requires more than reading texts or attending lectures. Meaningful and focused interactions are essential to helping students develop such knowledge. Fostering such interaction is no simple task.

Let us be honest: the nature of the topic makes it challenging. Aspects of culture and communication can be abstract, and cultural differences can manifest in unexpected ways. The limited and generalized nature of readings and lectures might not be the best method to foster learning about this topic. Rather, if students can review and interact with materials created by and for other cultures, they can begin to understand how cultural factors affect content expectations. If they can compare the content expectations of other cultures with their own, then they can better observe how culture guides content consumption. Additionally, if students can interact directly with individuals from other cultures, they can ask questions about and gain direct input on connections between culture and content consumption. All these factors are essential to understanding the complex and nuanced ways culture crafts content consumption. Instructors teaching culture and CM should thus have students review, analyze, and compare the content-related expectations of different cultural groups.

Raising awareness involves more than simple observation. Students need to actively interact with materials in informed or guided ways, for such directed activities can help them process how culture affects content consumption. They also need to discuss what they learn, in guided and meaningful ways, to process their experiences and observations in relation to their peers. Doing so can help them identify limitations in what they observe and understand as well as compare their experiences with others and shape new world views.

To address such factors, instructors should use small-group and individual activities requiring students to interact with content and with individuals from other cultures. Next, students would engage in guided, in-class discussions of these activities. These discussions would be sessions where students review materials and formulate and share related ideas, observations, and impressions with their peers. This also means instructors need to guide these discussions to help students understand how to conceptualize the tasks they have performed in relation to implications for CM practices. These discussions allow students to present and reflect on the results of their activities. They also involve the instructor helping students talk through/think through what these findings mean in terms of culture and CM. Such approaches can help students develop an understanding of how culture shapes content consumption. They can use this knowledge to avoid problematic assumptions and to guide future research when engaging in CM in cross-cultural contexts.

The key to this approach is instructors using activities that allow students to interact with content and individuals from other cultures in guided ways. These are assignments that task students with identifying expectations of content consumption in different cultures. Effective learning in this situation involves the nature of the activity and the materials or individuals selected for it. Instructors can use certain well-planned and highly focused exercises to achieve this goal. The next section provides examples of activities instructors can use to teach students about CM in cross-cultural contexts.

Assignments

The more students can interact with other cultures and review content other cultures create, the more aware they can become of how culture can affect CM practices. Instructors can use a variety of CM activities to help raise student awareness of such factors. Access to materials and individuals is central to such undertakings. So, too, is the structure of the related tasks once instructors locate appropriate resources.

This section provides examples of three kinds of activities instructors might use to familiarize students with cultural factors affecting CM practices. Each exercise represents an independent activity instructors can use when teaching this topic. Each of the three activities can also be combined with other activities described here to provide students with multiple opportunities to investigate culture and CM. These activities also provide different approaches based on

- Instructor familiarity with the topic.
- Access instructors (or their students) have to resources—such as international populations.

Instructors can select the activity that might work best depending on the nature of the class.

Activity 1: Comparative Online Analysis of Cultures

This activity can be done as an in-class exercise, a homework activity, or a mix of the two. For this activity, students review content designed for a specific audience from another culture. The objective is for them to identify patterns of what content is used and how it is organized and published via online media in a particular genre. Each group of students would use this activity to develop guidelines for how individuals from the culture studied might approach content in a given genre. To this end, this activity is a mix of individual, small-group, and overall class exercises that instructors can assign as a series of in-class activities and homework.

Preparation Work

To prepare for this exercise, the instructor needs to identify a particular sector/industry in which multiple organizations have an international presence (e.g., the auto industry, the soft drink industry). Next, the instructor would need to review the websites of two or three of the organizations in this industry to see if they have created websites for different international audiences (e.g., a website for China, for Brazil, for France). The goal is to create a listing of five to seven countries (depending on the size of the class) for which all organizations reviewed have created a website (e.g., Coca-Cola and Pepsi both have created an independent website for China, for France, and for Brazil). These websites will become the materials students analyze for this activity.

The Activity

For this activity, the instructor should divide the class into small groups of three or four students. Next, the instructor would ask each group to review the website a particular company uses to share information about its products and services with Anglo-American audiences. The groups should spend ten to 15 minutes reviewing this site and create a checklist of content-creation guidelines for

- The kinds of content used on the site (e.g., amount of visual vs verbal content).
- The formatting of that content (e.g., mostly two- to four-sentence paragraphs or in lists accompanied by images).
- The organization of the sites' content (e.g., how many images are on a page and where are they located; how is text organized on pages; where are menu bars and navigation features located).

Each group would next present its CM guide/checklist to the class, and the instructor would write suggestions on a whiteboard/central display area for all to see. Once all groups have presented, the class would discuss how to create a final CM guide for using websites to share information on the related company with an Anglo-American audience. (The instructor needs to make—or to have a student make—a copy of these final guidelines for future use.)

For the next part of this activity, the instructor would have students remain in their groups and assign each group the website that the related organization created for a different culture. (This list of cultures the instructor assigns would be based on the initial review of sites done prior to the class.) Each student team would then be assigned three associated tasks:

Task 1: Individual Review

Each team member needs to individually review the website for that assigned culture and develop CM guidelines for creating websites for members of that culture. Students would again focus on

- The kinds of content used on the site (e.g., amount of visual vs verbal content).
- The formatting of that content (e.g., mostly two- to four-sentence paragraphs or in lists accompanied by images).
- The organization of the sites' content (e.g., how many images are on a page and where are they located; how is text on a page organized; where are menu bars and navigation features located).

Students would have 15–20 minutes in class to complete this review, or it could be done as a homework assignment.

Note: this activity focuses on the visual design of the site; that is, how content appears to be presented vs what it says. It is therefore fine if students cannot read the language of the assigned site.

Task 2: Group Comparison

The second task has groups discussing the CM guidelines they developed for the culture they all reviewed. The objective is for the group to create one set of guidelines all members would use with CM practices for audiences from that culture. This discussion could be done as a ten- to 15-minute in-class activity or a homework assignment.

Task 3: In-class Presentation

Each group would select a spokesperson to share the group's CM guidelines to the overall class via in-class presentation. (Ideally, groups will have the ability to display the website they reviewed to the class as they present these guidelines.) After presentations end, the instructor would display the Anglo-American CM guidelines developed during the first class activity. The instructor would ask students to compare and contrast these original guidelines with the guidelines groups created for another culture. The objective is to help students realize CM guidelines designed for one culture cannot automatically be used to create content for another. During this discussion, the instructor would ask students how they might revise content designed for Anglo-American audiences to make it meet the content expectations of the culture they reviewed.

This final discussion helps familiarize students with how content expectations can vary across cultures. It can also help students realize how such variations can occur across different aspects of content in the same medium and for the same product or topic. Such comparative activities can help students understand the importance of considering culture when engaging in CM projects or globally distributed teams. Ideally, this knowledge will prompt students to reflect more carefully and seek out expert help when addressing CM practices involving other cultural groups.

Note: The online nature of this activity means it can be performed in a range of contexts and at any point in a given term as it is not dependent on access to or the schedules of other individuals.

Activity 2: Interviews with Individuals from Other Cultures

The more directly students can interact with other cultures, the better they can understand the content-related expectations members of a culture might have. Interviews with multiple individuals from another culture can provide students with such information. These interviews can also help students see how cultural factors affect perceptions and uses of content.

Preparation Work

For these activities, the instructor needs to identify three or four members of the same culture (one other than the students' native culture) who students might interview. The instructor would then contact these individuals and ask

- If they would let students interview them about how they perceive and use certain materials.
- What times they might be available for such interviews.

Ideally, these would be individuals who live and work locally so they could participate in on-site interviews with students. If not, instructors might seek out persons willing to do Skype-based interviews for this activity.

Key to this activity is that students compare responses across multiple individuals from a culture. Such comparison allows students to see if general patterns emerge in relation to content consumption. For this reason, at least three individuals from the same culture should agree to being interviewed. The goal is to provide input from multiple perspectives and identify similarities across all sources of information. (In essence, one person's statement represents a personal opinion, similarity between two persons could be coincidence, but overlap across three persons indicates a prospective trend.) Ideally, the more subjects there are to interview, the better. The instructors, however, will need to determine what parameters are workable based on the time available and the individuals to whom they have access.

The Activity

Due to the constraints of access to individuals, these interviews could take various forms:

Option 1: Comparative Individual Interviews

Ideally, the instructor could identify enough interview subjects so groups of students could meet with interviewees to ask genre-CM-related questions, such as:

- What kinds of materials do you use for doing X?
- What kinds of information do you expect to find in those materials?
- How do you expect those materials to be organized?
- Have you ever noticed anything different about how American materials are organized or the information they contain? Can you describe those differences?

In this situations, students would be divided into teams of three to five, and each team would meet with three or four interviewees from a given culture to ask these questions. Next, students would compare interviewee responses to look for patterns in how individuals from that culture might perceive and react to content. Students would use this comparison to develop guidelines for creating content for members of that culture. Each group would present its guidelines to the class for comparison and discuss how to create an overall set of guidelines for creating content for the related cultural group.

The objective of this discussion is to raise student awareness of how cultural factors affect expectations of content. It is also to help students understand how CM practices need to be re-considered, and new approaches used when developing content for different cultural audiences.

Option 2: Comparative In-class Interviews

It is possible only a limited number of individuals are available for interviews. In this case, the instructor should avoid taxing interviewees with multiple interviews by different groups of students. Instead, the instructor might invite the interviewee to visit the class—either in person, via Skype, or a mix of the two. During this visit, students would present their interview questions as a class. The interviewee could then respond for the entire class to hear, and one or more members of the class could be tasked with recording these responses. (These students might also be tasked with creating a related PowerPoint or other presentation of the responses from this interview session.)

If time permits, all three (or more) interviews might take place on the same day, or they might need to be scheduled for different class meetings during the term. Once all these interviews are completed, the instructor would display the results of/notes from each set interviews for all students to see and compare. The class would use these notes to discuss how to create CM guidelines for creating content for members of that culture.

As with Option 1, the objective of this activity and related discussion is to raise student awareness of how cultural factors affect expectations of content. It is also to help students understand how CM practices need to be re-considered, and new approaches used when developing content for different cultural audiences.

Note: The options described here represent two different approaches for doing these kinds of interviews. The goal, however, is for students to actually interact with individuals from other cultures. Instructors should consider how they might modify or adapt this approach to address the realities of the parameters and limitations of their course and its schedule.

Activity 3: Presentations by—and Interviews with—Localizers

As noted, localization plays an important role in international and cross-cultural CM. Students can benefit from opportunities to interact with and ask questions of localizers who work in CM. This activity is designed to facilitate such interactions.

Preparation Work

For this activity, the instructor would identify one or more localization professionals who work in the area of CM. Next, the instructor would contact this individual/these individuals about doing an on-site or Skype-based guest lecture on CM practices and localization. After invitees agree, the instructor should ask if they can recommend two or three sources on the topic to share with students prior to their lecture.

The Activity

Two weeks prior to the guest lecture, the students would be informed of/reminded that a guest lecturer form the localization industry will be meeting with the class on date X. To prepare for this meeting, students would need to identify three to five sources—articles, book chapters, or web pages—on CM and localization or on CM in international/global contexts. Students would submit (digitally) a list of these sources with a related one-paragraph summary of each. The instructor would compile these sources into a list (as well as add sources suggested by the invited guest lecturers) and assign students to read these entries in order to discuss them with the class before the invited lecture.

During this discussion, the instructor would task the class with developing a list of five to seven CM-related questions to ask the guest lecturer after her/his presentation. Sample questions the instructor could use to guide this discussion might include:

- What is localization?
- How is localization connected to CM practices?
- What are the most common localization challenges or problems for you in terms of CM practices?
- What suggestions do you have for what individuals should do to make content and CM practices easier in terms of localization?

The instructor would distribute the final, agreed-upon questions to students in advance of the guest lecture, and identify which student would ask each question. The instructor would also inform students they would be responsible for taking notes on the response to these questions. Students would use these responses and information in the localization readings the instructor assigned to draft a short report on global CM practices.

On the day of the guest lecture, the presenter would first share her/his perspectives on connections between localization and CM as well as suggestions for effective CM practices for global contexts. Next, the students would present the previously-agreed-upon questions and, if time permitted, the floor could be opened to additional questions. Students would then have a certain amount of time (e.g., seven to ten days) to use information from the guest lecture and the assigned localization readings to draft their paper on global CM practices. The instructor might ask students to submit this paper as a written report or present it in class, or both. In all cases, once all papers are submitted/ presentations are completed, the instructor should ask the class to discuss, compare, and contrast what they reported on. This discussion can help students gain a greater understanding of how to work with localizers to engage in effective CM practices in global contexts.

The objective of this overall activity is to raise student awareness of what localization is and how localization can be important to CM in global contexts. It is also to introduce students to the idea of—and practices for—working with localizers on future cross-cultural CM projects.

Conclusion

Navigating global and cross-cultural contexts is no easy task. Doing so, however, is increasingly essential to organizational success in the modern world. The distributed and single-sourced nature of CM positions it as a mechanism for adapting to these dynamic environments. Success in such situations involves understanding the different expectations audiences associate with content. An informed and well-planned approach to CM can effectively address such factors.

This chapter has provided an overview of how technical communication instructors can introduce students to factors of cultural expectations and CM practices. The key is to make students aware that not all cultures hold the same expectations of what content should be included in materials they use. This entry has also presented a number of exercises instructors can use to raise student awareness of these factors. The more students can learn about and understand such aspects, the better they can plan for and address cultural preferences when working in greater global contexts.

References

Barker, D. (2016). *Web content management: Systems, features, and best practices*. Sebastopol, CA: O'Reilly Media.

Batova, T., & Clark, D. (2015). The complexities of globalized content management. *Journal of Business and Technical Communication*, *29*(2), 221–235. doi:10.1177/ 1050651914562472.

Berkenkotter, C., & Huckin, T.H. (1995). *Genre knowledge in disciplinary communication: Cognition/culture/power.* Hillsdale, NJ: Lawrence Erlbaum.

Campbell, C.P. (1998). Rhetorical ethos: A bridge between high-context and low-context cultures? In S. Niemeier, C.P. Campbell, & R. Dirven (Eds.), *The cultural context in business communication* (pp. 31–47). Philadelphia, PA: John Benjamins.

Cohn, N., Taylor-Weiner, A., & Grossman, S. (2012). Framing attention in Japanese and American comics: Cross-cultural differences in attentional structure. *Frontiers in Psychology, 3*(349). Retrieved from www.frontiersin.org/articles/10.3389/fpsyg.2012.00349/full.

Collins, R. (2010). *Three myths of internet governance: Making sense of networks, governance and regulation.* Exeter: Intellect Books.

Diorio, S. (2001). Governance issues in content management. *ClickZ.* Retrieved from www.clickz.com/governance-issues-in-content-management/59983/.

Driskill, L. (1996). Collaborating across national and cultural borders. In D.C. Andrews (Ed.), *International dimensions of technical communication* (pp. 23–44). Alexandria, VA: Society for Technical Communication.

Esselink, B. (2000). *A practical guide to localization.* Philadelphia, PA: John Benjamins.

Grundy, P. (1998). Parallel texts and divergent cultures in Hong Kong: Implications for intercultural communication. In S. Niemeier, C.P. Campbell, & R. Dirven (Eds.), *The cultural context in business communication* (pp. 167–183). Philadelphia, PA: John Benjamins Publishing Company.

Halvorson, K. (2008). The discipline of content strategy. *A List Apart.* Retrieved from https://alistapart.com/article/thedisciplineofcontentstrategy/.

Hannon Hill Corporation. (2010). *Introduction to content management systems.* Atlanta, GA: Author.

Lieb, R. (2018). *Global content strategy: This is going to be big.* San Francisco, CA: Kaleido Insights.

Maiella, T. (2015). 6 tips to launching a global content strategy. *Skyword.* Retrieved from www.skyword.com/contentstandard/marketing/6-tips-to-launching-a-global-content-strategy/.

Milanovic, B. (2013). The economic causes of migration. The Globalist. Retrieved from https://www.theglobalist.com/economic-causes-migration/.

Miller, C.R., & Selzer, J. (1985). Special topics of argument in engineering reports. In L. Odell & D. Goswami (Eds.), *Writing in non-academic settings* (pp. 309–341). New York, NY: Guilford.

Optimizely. (2019). Content Management System.Retrieved from www.optimizely.com/optimization-glossary/content-management-system/.

Rockley, A. (2016). Why you need two types of content strategist. Content Marketing Institute. Retrieved from https://contentmarketinginstitute.com/2016/02/types-content-strategist/.

Romero, S. (2017). Spanish thrives in the U.S. despite an English-only drive. *New York Times.* Retrieved from www.nytimes.com/2017/08/23/us/spanish-language-united-states.html?emc=eta1.

St.Amant, K. (2005). A prototype theory approach to international web site analysis and design. *Technical Communication Quarterly, 14,* 73–91. doi:10.1207/s15427625tcq1401_6.

St.Amant, K. (2006). Globalizing rhetoric: Using rhetorical concepts to identify and analyze cultural expectations related to genres. *Hermes— Journal of Language and Communication Studies, 37,* 47–66.

St.Amant, K. (2015, April). Reconsidering social media for global contexts. *Intercom*, 16–18.

St.Amant, K., & Rice, R. (2015). Online writing in global contexts: Rethinking the nature of connections and communication in the age of international online media. *Computers and Composition*, *38*(B), 5–10.

Scime, E. (2009). The content strategist as digital curator. *A List Apart*. Retrieved from https://alistapart.com/article/content-strategist-as-digital-curator/.

Skyword. (2015). Skyword launches Skyword Global, giving enterprises the resources to deploy content marketing programs worldwide. Retrieved from www.skyword.com/skyword-launches-skyword-global-giving-enterprises-resources-deploy-content-marketing-programs-worldwide/.

TCBOK. (n.d.). Content management systems. Retrieved from www.tcbok.org/wiki/content-management-systems/.

Tebeaux, E. (1999). Designing written business communication along the shifting cultural continuum: The new face of Mexico. *Journal of Business and Technical Communication*, *13*, 49–85.

Ulijn, J.M. (1996). Translating the culture of technical documents: Some experimental evidence. In D.C. Andrews (Ed.), *International dimensions of technical communication* (pp. 69–86). Arlington, VA: Society for Technical Communication.

Ulijn, J.M., & Strother, J.B. (1995). *Communicating in business and technology: From psycholinguistic theory to international practice*. Frankfurt: Peter Lang.

Walzer, A.E. (2000). Aristotle on speaking "outside the subject": The special topics and rhetorical forums. In A.G. Gross & A.E. Walzer (Eds.), *Reading Aristotle's Rhetoric* (pp. 38–54). Carbondale, IL: Southern Illinois University Press.

Weiss, S.E. (1998). Negotiating with foreign business persons: An introduction for Americans with propositions on six cultures. In S. Niemeier, C.P. Campbell, and R. Dirven (Eds.), *The Cultural Context in Business Communication* (pp. 51–118). Philadelphia, PA: John Benjamins.

Woolever, K.R. (2001). Doing global business in the information age: Rhetorical contrasts in the business and technical professions. In C.G. Panetta (Ed.), *Contrastive rhetoric revisited and redefined* (pp. 47–64). Mahwah, NJ: Lawrence Erlbaum.

Yunker, J. (2003). *Beyond borders: Web globalization strategies*. Indianapolis, IN: New Riders.

AFTERWORD: BEYOND MANAGEMENT

Understanding the Many Forces that Shape Content Today

Carlos Evia

VIRGINIA TECH

Rebekka Andersen

UNIVERSITY OF CALIFORNIA, DAVIS

Chapter Takeaways

- Technical communication educators need to embrace content management as a core disciplinary competency.
- The term "content management" is a small umbrella to cover all content-related work. The field of technical communication should move beyond talking about content management to talking about content operations and the disciplines of content.
- Structured authoring is key to modern publication workflows and practices. Academia can no longer afford not to talk about structured content.
- Ongoing collaboration between academic programs and industry partners is essential to prepare future technical communication professionals for success in a Content 4.0 world and beyond.

Keeping abreast of trends in the content profession is something that we both try to do as much as possible. We attend at least one industry conference a year, such as the Center for Information-Development Management (CIDM) Best Practices Conference or the Society for Technical Communication (STC) Summit; regularly listen to podcasts; watch webinars; engage with influential practitioners in blogs and social media; and read trade publications. Carlos, in his role as Co-Chair of the Lightweight DITA (Darwin Information Typing Architecture) Subcommittee and member of the DITA Technical Committee at the Organization for the

Advancement of Structured Information Standards (OASIS), also engages in frequent conversations with committee members regarding standards development and implementation approaches. And Rebekka, as an academic member of the CIDM Advisory Council, hosts annual conference calls with CIDM members to learn about what their organizations are doing well and what challenges they are facing.

We have, in different ways, strived to shape our field's understanding of and approach to teaching content management—Carlos through his standards-development work (Evia, Eberlein, & Houser, 2018; Evia, 2019) and course innovations (Evia, Sharp, & Pérez-Quiñones, 2015; Evia & Priestley, 2016) and Rebekka through her comprehensive reviews of the trade and scholarly literature (Andersen & Batova, 2015; Batova & Andersen, 2017) and articulations of content strategy practices in industry (Andersen, 2014). But keeping abreast of trends—and, more importantly, making sense of what they mean for educating future technical communicators—is proving to be an increasingly difficult task.

As chapters in this collection show, many technical communication academics are in a non-stop game of trying to catch up with industry practices related to content. Whereas some have apparently thrown in the towel (Kimball, 2016, for example, suggests that we turn our focus away from preparing "strategic" to preparing "tactical" technical communicators), others have tried to keep up and adapt curricular approaches (e.g., Batova, 2018; Duin & Tham, 2018). One would think that the number of academics who keep a finger on the pulse of the content industry would be on the rise, but the opposite seems to be true (as evidenced by the near absence of any recent published research focused on content activities in organizations). A key reason for this, we think, is that talking about content in technical communication is complicated, and continues to get more complicated.

Take, for example, the concept of Content 4.0, which Joe Gollner (2016a) introduced at the 2016 CIDM Best Practices Conference in his talk "Are You Ready for Content 4.0? The Shape of Things to Come and How to Prepare for It." Gollner, known in the content industry for always being a step ahead of what is to come, compared the four evolutions of industrial innovation to what he saw as the four evolutions of content. Given changes in industry practices, from labor divisions and mechanization (Industry 1.0) to automation of manufacturing tasks and smart parts (Industry 4.0), Gollner suggested that a parallel evolution of content innovation was inevitable (Figure 12.1 depicts this evolution). Whereas traditional publishing characterized Content 1.0, when content and format were intertwined, single-source publishing characterized Content 2.0, when content and format were separated; Gollner noted that, as of 2016, much of the focus of the content industry had been directed at Content 2.0. However, Content 3.0, which Gollner called *Integrated Content*, was what a majority of the leaders and innovators in the room were trying to figure out how to do well; the goal being for content contributors (e.g., marketing, training, technical publications) to share rules governing structure and semantics in such a way

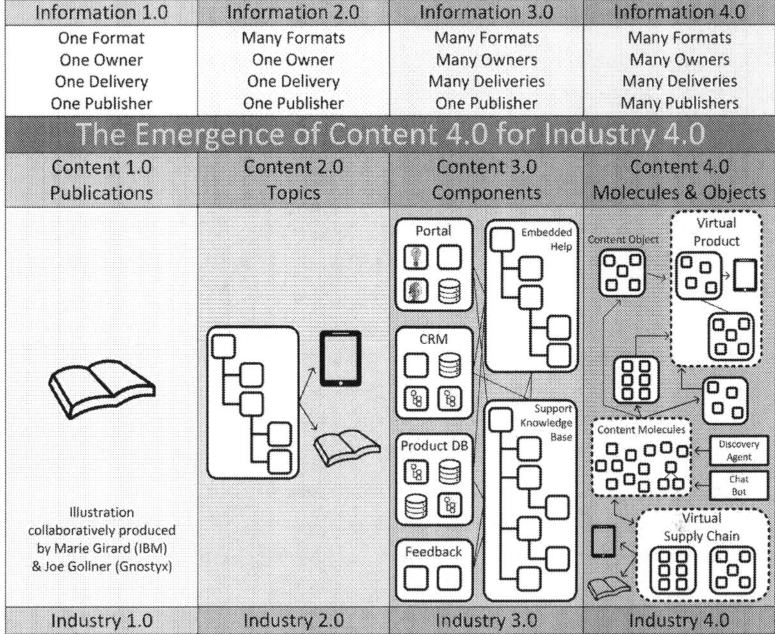

FIGURE 12.1 The emergence of Content 4.0 for Industry 1.0. In moving from Content 1.0 to Content 4.0, "the focus of communicators shifts from the publications themselves towards progressively smaller and smarter content components" and "there is a greater and greater awareness of, and interaction with, the various applications that govern either where the content sources hail from or where the content assets will go as part of their publication and delivery" (Gollner, 2016b).

Image produced by Marie Girard and Joe Gollner and used with permission

to allow for content integration and an increased focus on the customer or user experience journey.

Gollner then introduced Content 4.0, which he saw as the next stage of content innovation. Content 4.0, he suggested, would be all about smart content, or "content that assembles, transforms, and renders, dynamically in response to contexts and agent needs." In this evolution of content, the modular content object becomes the book or document of the Content 1.0 world:

> content is planned, designed, created, managed, and exchanged as objects that incorporate ... the associated rules governing the structure and meaning of the content and an array of rendition and behaviour processes that the object can use to render that material independently or in concert with other content objects.
>
> *(Gollner, 2016b)*

In a Content 4.0 world, "content becomes much more precise, content design becomes one part in a total system, content creation becomes more technical, and content management becomes more complex."

Rebekka, feeling in great need of Ibuprofen, was not the only one in the room at that moment staring perplexedly at Gollner, who looked apologetic. Technical communication had just entered the world of science fiction and, at least for those in the room who were not engineers and computer scientists, the path forward seemed daunting, if not impossible. Anticipating feelings of defeat, Gollner in his usual witty style projected an image of a face plant, then moved to assure those in the room that they were "headed in the right direction" and that "the burden of complexity can be shared" and "managed within engineered systems."

A few years earlier, at the same conference, the leaders and innovators in the room were talking about integrated content strategy as a concept. At that time, many companies had successfully adopted a single-source publishing model and were looking to further innovate. By 2017, many of the companies looking to implement an integrated content strategy had successfully done so (as evidenced by the number of success stories presented at the 2017 CIDM Best Practices Conference). A review of trending podcasts, webinars, blogs, and conference agendas now shows the content industry intently focused on the affordances of smart content, designed and engineered to interact with chatbots, voice assistants, and intelligent machines and to populate PDFs, online help, mobile, video, and other content delivery channels.

Recently, *content engineering* has emerged as a fully fledged discipline focused on the technical aspects of smart content publishing workflows, while the discipline of *content strategy* has matured in its focus on planning and measuring content assets and their impact on publications. Carlos tries to stay connected to the professional conversation on these emerging disciplines, and his assessment is that they are intrinsically connected to *content management* but each one has a unique edge. As we will describe later in this afterword, the combination of disciplines related to the planning, management, delivery, and assessment of technical content is a complex meeting of forces that cannot be grouped under the umbrella term of "content management" anymore.

When Tracy asked if we would be willing to contribute to this edited collection with an afterword that addressed the question "Where do we go from here?", we decided that we wanted to focus on the many forces shaping content today. We are presenting our overview of these forces as a way to lay some groundwork for thinking about what and how we might focus future conversations on the teaching of content management. In what follows, we describe some key forces shaping content activities in industry and make a case for why we, as educators, need to embrace content management as a core competency in technical communication. We also argue that the field needs to move beyond talking about content management to talking about content operations and the disciplines of content. Doing so would help academic programs to specialize curricular offerings and solve problems that already exist with modern technical content operations but that require a new vocabulary to be properly defined or explained.

The Many Forces Shaping Content

In this section, we explain what we see as the four primary forces shaping content conversations and activities in industry today.

The Need to Publish to Many Channels, Many Devices (Force 1)

User demand for content accessible from any device and available in an array of formats is perhaps the most significant force shaping content conversations and activities today. Traditional document-based publishing practices are not sufficient to meet this demand, and many organizations, in efforts to modernize their approaches to developing, managing, and delivering content, have reached out to consultant agencies specializing in content strategy and engineering for help. The challenge for organizations, as self-proclaimed content evangelist Scott Abel (2017) so aptly sums up in his personal bio, is figuring out how to "deliver the right content to the right audience, anywhere, anytime, and on any device."

Smart technologies and faster, better Wi-Fi are changing how and when people access content and what they expect from that content. In 2016, for the first time, more people accessed the internet via a mobile device than a computer (StatCounter, 2016). Gartner (2016) predicts that "by 2020, 30% of web browsing sessions will be done without a screen."

To compete in this technological landscape, organizations must now have a content presence across a range of channels and devices. And these devices do not just include computers, tablets, and smartphones. They include voice interfaces such as Alexa and Siri, smart wearables such as Apple Watch and Google Glass, and invisible interfaces. In a recent conversation with Carlos, Jack Molisani—a longtime prominent voice in the profession—had the following to say about the shape of content to come:

> We're going to get to the point where you have absolutely no idea where this content is going to be displayed. It could be displayed on a refrigerator, it could be displayed on a car dashboard—so you need to write this content so that it stands alone, can be easily found, and then dropped into whatever content ecosystem it's going to [eventually] show up in, which is so different than having to format something so that it can fit into an 8x11 page.

Results of three recent surveys of trends in technical communication reveal that PDF and HTML continue to be the most common publishing formats, with embedded user assistance and video close behind. Although fewer than

20% of organizations reported publishing to mobile applications, dynamic delivery, or chatbots, many had plans to publish to these formats in the near future. The surveys we reference here include the 2017–2018 edition of the Adobe Tech Comm Survey (2,000-plus respondents) (Singh, 2018), the 2018 Publishing Trends survey conducted by Comtech and the Data Conversion Laboratory (DCL) (340-plus respondents) (Stevens & Madison, 2018), and the 2019 State of Technical Communication survey conducted by the Content Wrangler (600-plus respondents) (Abel, 2018a). Table 12.1 presents a snapshot of the range of channels to which organizations are currently publishing content and to which organizations plan to publish content in the near future, according to the results of the three surveys.

The challenges that organizations face in preparing content for a multichannel publishing world are many. These range from the need to adopt strategies, processes, and technologies that enable the creation of highly engineered, modular content to the need to work across departmental units and teams to develop unified processes that allow for content to be effectively shared and combined in myriad ways.

In a world where people access product content from a vast array of channels and devices, that content—regardless of where or how it is accessed—needs to

TABLE 12.1 Current and Future Channels to Which Organizations Are Publishing Content

Channel	*Currently Publishing*	*Prediction for Near Future*	*Noteworthy*
PDF	75% (Comtech/DCL) 86% (Adobe)	20% decline –	PDF publishing predicted to decline
HTML	80% (Comtech/DCL) 36% (Adobe) 56% (Content Wrangler)	Same 17% increase –	HTML5 is the format choice for mobile, with 54% of Adobe respondents reporting use of HTML5 for this
Embedded User Assistance	48% (Comtech/DCL) 15% (Adobe)	4% increase 3% increase	
Mobile Apps	18% (Comtech/DCL) 16% (Adobe)	14% increase 13% increase	
Video	15% (Comtech/DCL) 66% (Content Wrangler)	8% increase 18% increase	
Dynamic Delivery	20% (Comtech/DCL)	30% increase	Dynamic delivery predicted to increase significantly
Chatbots	6% (Adobe)	22% increase	62% of Content Wrangler respondents said they would be launching chatbots within six months

communicate a consistent, coherent message. How to do this well is one of the key challenges content leaders are working to overcome today. We call this challenge Force 2, the need to provide seamless content experiences.

The Need to Provide Seamless Content Experiences (Force 2)

Not only do users increasingly expect to be able to access the content they need any time, from any device, in the format they need it (e.g., PDF, web, video), they also expect (and need) a consistent, unified message for the same type of content (e.g., product description) regardless of where or how it is accessed. When the same type of content is written in different ways and information conflicts, customers or users can easily become confused and frustrated. Thus, one of the strongest forces shaping content today is the need for organizations to provide seamless content experiences to users or consumers.

Cruce Saunders (2018), a content engineer and founder of the content services agency [A], describes this need as follows:

> Today's thriving enterprises recognize the value of content and the role it plays in the customer experience. Consumers now expect highly relevant, timely, personalized content at all touchpoints of the customer journey, on any device or interface, across any and all channels of their choosing. Organizations that successfully deliver on these expectations reap the financial rewards and become leaders in their industries, while those that do not struggle to keep up with the rapid rate of change we all see around us.
>
> *(p. 9)*

But creating seamless content experiences is a massive undertaking for any organization. Saunders, this time in an interview with Scott Abel published in *Intercom* (2018b), points to the "soup of existing content systems, structures, and standards that don't connect and communicate" (p. 22) as the biggest obstacle standing in the way. He suggests that "we have too many content authoring management and publishing systems trying to orchestrate a single customer experience with content that doesn't transit between systems, is redundant, or is otherwise overwhelmingly inefficient" (p. 26). Part of the problem is that organizations have tended to invest more in content delivery than in content production. They have been intently focused on engaging users through multiple channels, or the multichannel experience. Angus Edwardson (2019) suggests that this focus has become problematic for customers because although they can engage with content through various channels and platforms, they "still lack a seamless experience and consistent messaging across each of these channels."

The concept of omnichannel delivery has emerged in the past couple of years to help organizations think through how to orchestrate seamless content experiences. Whereas multichannel is focused on the processes and technologies for delivering content across an array of channels and devices, omnichannel is focused on coordinating activities across departmental silos to achieve a consistent message to users wherever they interact with an organization (OmnichannelX, 2018). It is concerned with addressing the very problem that Saunders articulates.

At the core of an omnichannel initiative are the disciplines of content strategy and content engineering. We talk more about these disciplines later in the section titled Beyond Content Management.

The Need to Redefine Our Relationship with Subject Matter Experts (Force 3)

A disciplinary pillar of technical communication is the claim (sometimes implicit, but at other times quite explicit) that subject matter experts (SMEs) cannot write and therefore need us. This "us vs them" schema has been active for decades, and some have tried to study it. Mallette and Gehrke (2018), for example, point out that despite the complex reality of their work assignments and the dynamic role they can play, "technical communicators are often still confined to the box of documentation-ist, rather than the more elevated status of subject matter expert … in the various subjects in which they work" (p. 75).

In the early 1990s, Walkowski (1991) conducted a survey of software engineers to document the qualities that they most and least appreciated in a technical writer. Her findings showed a wide gap separating "us" from "them," as stated in the following quote:

> Engineers appreciate working with writers who enjoy being writers. Several engineers complained of writers who secretly (or not so secretly) want to be engineers and compete with them at every turn.
>
> *(p. 66)*

In a complementary study, Lee and Mehlenbacher (2000) surveyed technical writer professionals to address their perceptions toward SMEs. In their study, Lee and Mehlenbacher found that technical writers resented the stereotype that presented them as "glorified secretaries" of the SMEs.

In contemporary content operations, the writing expectations of technical writers and SMEs definitely challenge the mutually exclusive model of decades past. This change is particularly evident when working with application programming interface (API) reference documentation, which Angelini (2018) explains as follows:

> Web service APIs … are code frameworks widely used to offer a huge variety of services and information on the web, making them a strategic asset of web companies (like Google, Facebook, and Amazon). In order

> to integrate web services into their web sites and applications, web developers need a specific type of documentation called 'API reference', which is written following a predefined template.
>
> *(p. 70)*

The need to document APIs often resides in a communication structure that has SMEs taking the roles of authors and readers. In such a structure, a technical communicator needs to understand the product with the same depth as engineers, or an SME needs to be able to follow a style guide and template to create content that compares to what technical communicators can produce.

Influential blogger Tom Johnson, on his blog "I'd Rather be Writing," has written about this trend quite a bit over the past year, claiming that technical communication "has undergone a major evolution from end-users to developers as the target audience" (Johnson, 2018a, para 44) and that "developer tools and languages are becoming increasingly specialized and complex, requiring engineers to play more active, collaborative roles in documentation" (Johnson, 2018b, para 1). Content strategist Keith Schengili-Roberts (2018), who blogs with the pen name of "DITA Writer," notes that as work cycles get shorter and methodologies like Agile take hold, the SME has increasingly become a direct contributor to the process of delivering technical content

A consequence of this force is that content standards like DITA, which were traditionally reserved for technical communicators, now make it easier for SMEs to contribute to content repositories. As Angelini (2018) and others point out, SMEs can write reference topics according to a predefined structure and review topics for quick approval before publishing. However, SMEs used to working in lightweight programming languages can perceive learning DITA or another XML (Extensible Markup Language) grammar as an impediment to getting work done, and prefer to work with simple text formats like Markdown. Lightweight DITA (LwDITA) allows authors to do just that. In his book *Creating Intelligent Content with Lightweight DITA*, Carlos describes LwDITA as "a simplified schema for structuring content, with fewer elements, tighter content models and a simplified specialization architecture to define new types compared to those of DITA XML" (Evia, 2019, p. 16). LwDITA content can be authored in three different compatible formats:

- XDITA, based on DITA XML.
- HDITA, based on HTML5.
- MDITA, based on Markdown.

SMEs and technical authors can create different types of content using these compatible formats. Content topics created in these formats can be easily integrated into deliverables that will not reveal their source language to audiences of developers or end users. Content created in these collaborative workflows most likely will not be stored in isolated desktop publishing files, but instead will be shared in repositories open for distributed authoring and editing, which are at the center of the next force.

The Need for Continuous Content Development (Force 4)

For decades, the "us vs them" paradigm of SME-technical writer interaction presented in the previous force was maintained by a model of production known as waterfall. A product (physical or digital) was developed in a linear structure and "technical writers' work began at the end of the waterfall process, as products were rolling off the line" (Baehr, 2019, p. 6). Technical writers had to learn how to use the product (frequently based on a prototype or spec) and then document it in a manual for end users. If the product was defective, or if the writer made a typo or spelling error, the user manual would not change until a new version was produced through the waterfall.

A group of "organizational anarchists" (Highsmith, 2001) published *The Agile Manifesto* in 2001 "to contribute to a human-focused turn in how contemporary software development teams are managed" (Lauren, 2018, p. 29). Agile development models emphasize "teamwork, working deliverables, customer collaboration, and responsiveness to change." The Manifesto's principles "are called *adaptive* (or sometimes called *agile*), in contrast to waterfall development's *predictive* product development cycle" (Baehr, 2019, p. 7, emphasis in original). As software developers adopted Agile methods and approaches, the blurry lines between SMEs and technical writers enabled the adoption of Agile in the content development profession.

Primarily in content workflows in which the main deliverable is aimed for online distribution, many technical writers have embraced Agile practices that enable them to treat content components as computer code. Professional communities online and in real life promote the docs-as-code or docs-like-code movement (where *docs* is short for *documentation*). Anne Gentle (2017) lists the following criteria for authors involved in projects treating docs like code:

- Store the doc source files in a version control system.
- Build the doc artifacts automatically.
- Ensure that a trusted set of reviewers meticulously reviews the docs.
- Publish the artifacts without much human intervention.

An overall objective of the docs-as-code movement, which strengthens the previous force, is to enable "a culture where writers and developers both feel ownership of documentation, and work together to make it as good as possible" (Holscher, 2017). Unlike the waterfall environment that saved bug fixes or content corrections for major releases of printed documentation, the docs-as-code culture promotes an environment of continuous integration ("code is continuously tested, integrated with other code changes, and merged") and continuous development ("code is continuously deployed with each patch to the entire code base") (Gentle, 2017).

In practical terms, this approach to content development, management, and deployment requires a collaborative version control system like GitHub, and a content authoring format that resembles contemporary programming languages. This force represents a challenge for technical communicators used to working in a heavily structured language such as DITA. Practitioners involved in docs-as-code workflows lament that "whereas programming and scripting languages move toward simplified syntax and tagging systems, technical communication continues to rely on XML and complex, nested tag structures" (Evia & Priestley, 2016, p. 26). The tendency to use lightweight programming and markup languages to replace more powerful (and thus more verbose) structures like XML is also modifying the key enabler of the forces shaping content, which we discuss in the next section.

Enabling the Forces: Structured Content Is Key

In the previous section, we presented a brief overview of the primary forces shaping content conversations and activities today. The forces that we have identified in this afterword come from our interaction with the academic and practitioner sides of technical communication and are by no means authoritative or definitive. However, these forces are brought on by rapidly maturing and proliferating technologies transforming our profession and, as a result, require modifications and updates in the curricula of our academic programs.

A transformational effect of these forces that we can already see in the workplace is the need for organizations to create well-structured, semantically-rich content that is human and machine readable. Rob Hanna (2018), for example, has argued that to support seamless content experiences, technical communication practice will demand "greater precision, knowledge, and discipline in how we create content across an omnichannel universe." Content will need to become more granular and contain richer metadata, and each content container (or topic) will need to contain even smaller blocks (elements) of reusable content. The type of content that can be stored in chunks to be assembled based on the needs of a specific audience or context requires a process of structured authoring, which has been defined as "a publishing workflow that lets you define and enforce consistent organization of information in documents, whether printed or online" (O'Keefe & Pringle, 2017, p. 2). We believe that this type of authoring is key to organizations' ability to publish content to many channels and devices, provide seamless customer or user experiences, facilitate collaborative authoring between SMEs and technical communicators, and support continuous content development.

Structured authoring requires the separation of content and presentation, which "can create philosophical and cognitive dissonance for technical communicators trained to think of information as content that is inherently linked to presentation" (Clark, 2008, p. 36). That separation allows the creation of content "that is planned, developed, and connected outside an interface so that it's ready for any interface" (Atherton & Hane, 2018, p. 32).

In academia, we can no longer afford not to talk about structured content. In its annual survey of over 2,000 self-identifying technical communicators, Adobe (2018) found that adoption of structured content in industry had reached the halfway point, growing from 20% to 50% between 2012 and 2017. This finding suggests that depending on a word processor or desktop publishing tool that combines content and presentation is no longer a workable solution for technical communicators unless the only deliverable to publish is a single PDF document.

In contemporary content workflows, however, not all structured authoring projects look the same. Although some projects in heavily regulated industries or manufacturing environments do require the constraints and full reuse capabilities of DITA XML described by Jason Swarts earlier in this collection, web-aimed content created in Agile processes can benefit from a lightweight language like Markdown. That is the purpose of MDITA, the Markdown-based authoring format of LwDITA, which Carlos introduces and explains in his book *Creating Intelligent Content with Lightweight DITA*.

Beyond Content Management

Now that we have described the forces shaping content conversations and activities in industry and academia, this section will connect those forces to the present and future of content management as a term and object of study.

Adopting a More Precise Definition of Content Management

The term "content management" is a small umbrella to cover all content-related work. In academic conversations and publications in general, we have used *content management* to describe a wide variety of activities related to publication in digital environments. In this collection, Saul Carliner (Chapter 2) analyzes the complicated relationship between content management and technical communication. Carliner points out that technical communicators do not have a single identity and that jobs in our field vary widely. The same can be said about *content management* as a concept, and as a field we need to be more precise in how we talk about content activities. This content misnomer also affects industry conversations and, most importantly, job postings. Jokingly, Carlos and some technical content consultant friends have started talking about the *cosa nostra del contenuti* or "this content thing of ours" because *content management* or even *component content management* do not really explain all the activities and processes behind the planning, creation, publication, and revision of content.

In industry and academia, content management comes in web content management, component content management, enterprise content management, and many other flavors depending on the specific discipline or even organization using the term. Deane Barker (2016a) addresses this naming problem from the perspective of web content management. He defines *content* as "information produced through editorial process and ultimately intended for human consumption via publication" (p. 5). Barker adds that this definition "points to a core dichotomy of content management: the difference between (1) management and (2) delivery." He claims that these two disciplines require different skills and mindsets, and that "the state of current technology is creating more and more differences every day" (p. 5).

Although Barker only mentioned two disciplines, recently "content management" has been used as a general term to describe job titles and activities related to content strategy, content engineering, content development, and bona fide content management. We thus argue for a move beyond content management to a more unified understanding of the many forces that shape content today.

Bringing Content Forces and Activities into Conversation

To take under consideration the forces we have presented in this afterword and to provide a more specific definition of job titles and activities related to content, we propose two approaches that have been used in professional conversations. One places content-related activities under the banner of "Content Operations" (CO, or ContentOps). Content operations is "the behind the scenes work of managing content activities as effectively as possible" with a mix of "elements related to people, process, and technology" (Jones, 2019, p. 162). Therefore, content operations is the "glue" between the (1) plan for content, and (2) the content management system in which it is managed and delivered (Barker, 2016b).

Another, more granular approach calls the "disciplines of content" by specific names, which include:

- *Content strategy* (the "what"), which defines strategic direction and focuses on the plan, the vision.
- *Content engineering* (the "how," "when," and for "whom"), which defines content structure, metadata, content reuse planning, taxonomy, and other content relationships.
- *Content management* (managing content after it has been created), which defines the operational processes supporting the content lifecycle, including policies, workflows, permissions, and editorial activities as supported by a content management system (Saunders, 2015).

As developers of curriculum at the course and program level, we need to think about roles and specialized skills associated with the different disciplines related to content beyond management.

Looking Forward

Miles Kimball (2016), in "The Golden Age of Technical Communication," characterizes the move to content management as the *Glass Age* of technical communication, the essence of which is the separation of content from form. This separation, he suggests, necessarily also separates strategic design from writing. Whereas technical writers in Content 1.0 contexts drew on their specialized knowledge and rhetorical expertise to decide how best to design, organize, and write documents, technical writers in post-Content 1.0 contexts "no longer design an integrated document in which they strategically apply visual and lexical rhetoric to solve human problems" (p. 9). Rather, they perform the "mundane task" of writing content chunks in accordance with standards and architectures designed by those in strategic design roles.

Kimball (2016) suggests that those in strategic design roles comprise a "smallish group of people paid well to think strategically about design, rhetoric, visualization, presentation, information architecture, technology, and usability" and those in technical writing roles comprise a "much larger, less-well-paid group of people who write fragmented paragraphs that they save to a database, never knowing exactly where or how they will be used" (p. 10). Although this description arguably underplays the rhetorical decisions that technical writers still must make in any communication context, the description does point to a decline of rhetorical agency for those in more traditional writer roles.

The authors in this collection provide ideas and examples that technical communication educators can use to prepare students for these more strategic roles that will, we hope, afford them meaningful agency in the profession. Carliner, for example, encourages educators to approach content management not as a technology but as an aggregate of asset management processes, editorial processes, reuse workflows, and various other system and human activities. Doing so puts the focus on the contributions of different roles (e.g., planner/designer, creator, production specialist, manager) engaged with content management and on developing curriculum aligned with those roles. Likewise, Hart-Davidson and Lauren (Chapter 10) offer the concept of "writing stewardship" as a way to understand the important but often invisible work of technical writers in organizations doing content management.

We would be remiss in this afterword if we did not address the topic of faculty training. Although this collection offers a variety of tested approaches for teaching content management to students (e.g., Potts and Gonzalez's assignments for web content strategy in Chapter 3, and Gesteland's introduction to XML in Chapter 6), it does not have the purpose of replacing a complete curriculum to prepare faculty to be competent teachers of content management. So how can faculty gain the requisite skills and knowledge to advance content management curricula? How can they do this with the limited time, resources, and institutional support available to them? Some ideas include the following:

- *Participate in a content management learning community for faculty.* We imagine organizations such as the Council for Programs in Technical and Scientific Communication (CPTSC) or Association of Teachers of Technical Writing (ATTW) sponsoring learning communities for faculty interested in developing skills and competencies associated with strategic design roles in the landscape of content operations. These communities would, above all, support learning, and perhaps might be viewed as virtual makerspaces, where members collaboratively problem-solve as they make content things.
- *Complete free online training.* Scriptorium, for example, hosts a series of free DITA courses at learningdita.com.
- *Participate in industry webinars, workshops, and online communities.* The Content Wrangler, the Data Conversion Laboratory, and Scriptorium offer free and frequent webinars focused on content management topics. Comtech Services and STC offer regular workshops, though at a cost. The Write the Docs community is active and welcoming and focused on improving the art and science of documentation.
- *Collaborate with industry organizations.* Academic organizations, such as ATTW and CPTSC, might collaborate with industry organizations such as CIDM to establish a professional development exchange program. Academic organizations could offer workshops and webinars on topics of interest to industry peers, such as research methods; in exchange, industry organizations could offer workshops and webinars to academic peers on a variety of topics pertinent to the content disciplines.
- *Seek out teaching-focused content management scholarship.* This collection should be at the top of the list, of course. But there are many additional resources, such as *Creating Intelligent Content with Lightweight DITA* (by Carlos Evia) and "Leveraging Industry Onboarding Materials in the Curriculum" (by Stan Doherty, 2017).

Along with faculty training, our field in general and programs in particular need to think about updating pedagogy courses for technical and professional communication graduate students interested in an academic teaching career. The composition pedagogy courses that the majority of programs require graduate students to take are not sufficient, often leaving those hired to teach and contribute to technical communication curricula unprepared. Some existing technical communication pedagogy courses are focused on preparing graduate students to teach the technical writing service course for an audience of science or engineering majors. The genre-based approach of many of those courses, however, can be antithetical to the agnostic content mantra of contemporary content operations.

And finally, looking forward, we believe that ongoing collaboration between academic programs and industry partners will be essential to preparing future technical communication professionals for success in a Content 4.0 world and beyond. Internship and mentorship programs will continue to be central to this preparation. Such experiential learning programs provide opportunities for students to apply what they are learning in the classroom to workplace situations and to gain hands-on experience with the processes, standards, and tools of the trade. Pairing students with more experienced technical communicators through mentorship programs also affords students potentially rich learning opportunities, networking opportunities, and career guidance. Creating advisory boards, too, can help shape program and curricular development, particularly in the areas of structure authoring, content strategy, and content engineering. The Technical Communication Advisory Board at the University of Minnesota Twin Cities serves as a model for how industry partners can contribute to content management competency development at the curricular and program levels. See Duin and Tham (2018) for a rich description of the board and how its members contribute.

In addition to inviting industry partners to serve as internship site supervisors, mentors, and advisory board members, programs might consider inviting them to share onboarding and other training materials designed for new hires. These materials might then be adapted for courses and for faculty development. Stan Doherty's Project Alcuin, an open-source repository of onboarding material for academic technical communication programs, is a great place to start (see Doherty, 2017). The repository[1] includes sample content in multiple source formats and architectures and a series of exercises focused on linear and modular writing, content and markup, content maintenance, content reuse, and intelligent content assembly.

Conclusion

Positioned on a bridge that connects information technology to the humanities, technical communication can rightfully claim credit for many episodes of humanizing technology that have made the world a better place. At the same time, that connection also carries the responsibility of constant learning. Whereas some of our neighbors in fields more traditionally grounded in the humanities can successfully teach the same text for decades, faculty in technical communication need to constantly update their syllabi and research interests.

1 See https://github.com/StanDoherty/project-alcuin.

An easy first step is to keep a finger on the discipline's pulse through social media and academic conferences, but more advanced levels of involvement include pursuing, as much as possible, continuing education opportunities, and seeking input from alumni and practitioners on course and curriculum design.

Some feel that this call for constant learning perpetuates a stereotype of the academic side of technical communication as subservient to the needs of industry. We believe that a harsh reality is that the relevance and sustainability of our field will depend, in part, on the extent to which our students are prepared for roles that afford them meaningful agency within the landscape of content operations. Additionally, a healthy relationship with industry based on reciprocal understanding and collaboration can enhance research agendas to support creativity and innovation for the next chapters of our field. And in one of those future chapters, we will redefine content management as a term, practice, and object of study.

References

Abel, S. (2017). Author: Scott Abel. Retrieved June 26, 2019, from https://thecontentwrangler.com/author/scottabel/#.

Abel, S. (2018a, December 13). The state of technical communication 2019 [Webinar]. Retrieved June 26, 2019, from www.brighttalk.com/webcast/9273/338293/the-state-of-technical-communication-2019.

Abel, S. (2018b). An interview with Cruce Saunders. *Intercom, 65*(5), 22–26.

Adobe. (2018). Adobe Tech Comm Survey 2017–2018 Findings. Retrieved November 21, 2018, from www.adobe.com/mena_en/products/technicalcommunicationsuite/whitepapers/adobe_tech_comm_survey_2017-2018_findings.html.

Andersen, R. (2014). Rhetorical work in the age of content management: Implications for the field of technical communication. *Journal of Business and Technical Communication, 28*(2), 115–157.

Andersen, R., & Batova, T. (2015). The current state of component content management: An integrative literature review. *IEEE Transactions on Professional Communication, 58*(3), 247–270.

Angelini, G. (2018) Current practices in web API documentation. *Proceedings of the European Academic Colloquium on Technical Communication*, 70–85.

Atherton, M., & Hane, C. (2018). *Designing connected content: Plan and model digital products for today and tomorrow*. San Francisco, CA: New Riders.

Baehr, C. (2019). *The agile communicator: Principles and practices in technical communication.* Dubuque, IA: Kendall Hunt.

Barker, D. (2016a). *Web content management: Systems, features, and best practices*. Sebastopol, CA: O'Reilly.

Barker, D. (2016b, January 27). The need for content operations. Retrieved June 26, 2019, from https://gadgetopia.com/post/9307.

Batova, T. (2018). Global technical communication in 7.5 weeks online: Combining industry and academic perspectives. *IEEE Transactions on Professional Communication, 61*(3), 311–329.

Batova, T., & Andersen, R. (2017). A systematic literature review of changes in roles/skills in component content management environments and implications for education. *Technical Communication Quarterly, 26*(2), 2017, 1–28.

Clark, D. (2008). Content management and the separation of presentation and content. *Technical Communication Quarterly, 17*(1), 35–60.

Doherty, S. (2017). Leveraging industry onboarding materials in the curriculum. *Proceedings of the 35th ACM International Conference on the Design of Communication – SIGDOC 17, 5 pp.*

Duin, A.H., & Tham, J. (2018). Cultivating code literacy: A case study of course redesign through advisory board engagement. *Communication Design Quarterly, 6*(3), 44–58.

Edwardson, A. (2019, February 13). Content operations? How to collaborate on content (effectively). Retrieved March 1, 2019, from www.boye-co.com/blog/2019/2/13/content-operations-how-to-collaborate-on-content-effectively.

Evia, C. (2019). *Creating intelligent content with Lightweight DITA*. New York, NY: Routledge.

Evia, C., Eberlein, K., & Houser, A. (2018). *Lightweight DITA: An introduction*. Version 1.0. OASIS.

Evia, C., & Priestley, M. (2016). Structured authoring without XML: Evaluating Lightweight DITA for technical documentation. *Technical Communication, 63*(1), 23–37.

Evia, C., Sharp, M.R., & Pérez-Quiñones, M. A. (2015). Teaching structured authoring and DITA through rhetorical and computational thinking. *IEEE Transactions on Professional Communication, 58*(3), 328–343.

Gartner (2016, October 18). Smarter with Gartner. Retrieved February 2, 2019, from www.gartner.com/smarterwithgartner/gartner-predicts-a-virtual-world-of-exponential-change.

Gentle, A. (2017). *Docs like code: Write, review, test, merge, build, deploy, repeat*. Austin, TX: Just Write Click.

Gollner, J. (2016b, December 28). Content 4.0. Retrieved June 26, 2019, from www.gollner.ca/2016/12/content_4-0.html.

Gollner, J. (2016a, September). "Are you ready for Content 4.0? The shape of things to come and how to prepare for it." Paper presented at the CIDM Best Practices Conference, Santa Fe, NM.

Hanna, R. (2018, April). "Preparing content for intelligent machines." Paper presented at the CIDM CMS/DITA North America Conference, Denver, CO.

Highsmith, J. (2001). History: The Agile Manifesto. Retrieved June 26, 2019, from https://agilemanifesto.org/history.html.

Holscher, E. (2017). Docs as code. Retrieved June 26, 2019, from www.writethedocs.org/guide/docs-as-code/.

Johnson, T. (2018a, October 2). Tech comm trends: Providing value as a generalist in a sea of specialists (Part II). Retrieved June 26, 2019, from https://idratherbewriting.com/2018/10/02/providing-value-as-generalists-in-specialist-contexts-part-2/.

Johnson, T. (2018b, October 2). Tech comm trends: Providing value as a generalist in a sea of specialists (Part I). Retrieved June 26, 2019, from https://idratherbewriting.com/2018/10/02/providing-value-as-generalists-in-specialist-contexts-part-1/.

Jones, C. (2019). *The content advantage [Clout 2.0]*. San Francisco, CA: New Riders.

Kimball, M.A. (2016). The golden age of technical communication. *Journal of Technical Writing and Communication, 47*(3), 330–358.

Lauren, B. (2018). *Communicating project management: A participatory rhetoric for development teams*. New York, NY: Routledge.

Lee, M.F., & Mehlenbacher, B. (2000). Technical writer/subject-matter expert interaction: The writer's perspective, the organizational challenge. *Technical Communication, 47*(4), 544–552.

Mallette, J.C., & Gehrke, M. (2018). Theory to practice: Negotiating expertise for new technical communicators. *Communication Design Quarterly, 6*(3), 74–83.

O'Keefe, S., & Pringle, A. (2017, April 12). *Structured authoring and XML*. Research Triangle Park, NC: Scriptorium Publishing Services. Retrieved May 17, 2019, from www.scriptorium.com/structure.pdf.

OmnichannelX Conference. (2018). Retrieved June 26, 2019, from https://omnichannelx.digital/.

Saunders, C. (2015). *Content engineering for a multi-channel world: Enabling the next generation of customer experiences*. Austin, TX: Simple A.

Saunders, C. (2018). A new content order for the multi-channel, multi-modal world. *Intercom, 64*(1), 9–11.

Schengili-Roberts, K. (October, 2018). *The rise of the SME within technical communications*. Paper presented at the LavaCon conference, New Orleans, LA.

Singh, A. (2018, March 21). Adobe Tech Comm Survey 2017–2018 [Webinar]. Retrieved June 26, 2019, from https://techcomm-trends-2018.meetus.adobeevents.com/.

StatCounter. (2016, November 1). Mobile and tablet internet usage exceeds desktop for first time worldwide. Retrieved June 26, 2019, from http://gs.statcounter.com/press/mobile-and-tablet-internet-usage-exceeds-desktop-for-first-time-worldwide.

Stevens, D., & Madison, K. (2018). 2018 publishing trends. *CIDM Best Practices Newsletter, 20*(4), August, 71, 74–81.

Walkowski, D. (1991). Working successfully with technical experts—from *their* perspective. *Technical Communication, 38*(1), 65–67.

INDEX